THE
MIDDLE EAST
STRATEGIC BALANCE
2004–2005

THE
MIDDLE EAST
STRATEGIC BALANCE
2004–2005

Edited by

Zvi Shtauber and Yiftah S. Shapir

sussex
ACADEMIC
PRESS

BRIGHTON • PORTLAND

JAFFEE CENTER FOR
STRATEGIC STUDIES

2 4 6 8 10 9 7 5 3 1

First published in 2006 in Great Britain by
SUSSEX ACADEMIC PRESS
P.O. Box 2950
Brighton BN2 5SP

and in the United States of America by
SUSSEX ACADEMIC PRESS
920 NE 58th Ave.   Suite 300
Portland, Oregon 97213-3786

British Library Cataloguing in Publication Data
A CIP catalogue record for this book is available from the British Library.

Library of Congress Cataloging-in-Publication Data
The Middle East strategic balance, 2004–2005 / edited by Zvi Shtauber and Yiftah S. Shapir.
    p. cm.
Includes bibliographical references.
ISBN 1-84519-107-2 (h/b : alk. paper) — ISBN 1-84519-108-0 (p/b : alk. paper)
    1. Middle East—Strategic aspects. 2. Middle East—Armed Forces.
    3. Middle East—Military policy. 4. World politics—21st century.
    I. Shtauber, Zvi. II. Shapir, Yiftah.

UA832.M524 2006
355'.033056—dc22
                                                                2005024379

Typeset and designed by G&G Editorial, Brighton & Eastbourne.
Printed by TJ International, Padstow, Cornwall.
This book is printed on acid-free paper.

Contents

Contents

Charts

Introduction

I

Protest, violence, terrorism, and instability joined the expectation of change and nascent signs of political transformation in the Middle East of 2004–5. Indeed, the status quo and longstanding dominant order of the Middle East were tackled by significant challenges that bore a different stamp than previous contests. During the past year, as Mark Heller observes in the opening chapter of this collection, "perhaps for the first time in modern history the anti-establishment forces included not just radical elements, but also pro-democracy forces."

Feeding the instability throughout the region was the concept of political transformation. The Bush administration, since late 2004 newly armed with the legitimacy and determination conferred by reelection, infused this concept into the political discourse of the Arab world. At the core of the concept lies the conviction that the war on terrorism will not succeed unless accompanied by profound transformations in the political culture of the Middle East. Therefore, local actors were forced to reposition themselves along new, more auspicious political spectrums. The preservation of stability was no longer the declared ultimate aim for advancing American interests, although on a practical level this aim was no less important than it had been in the past. Instead the goal became to reshape the Middle East through democratization in the broad sense of the term, which entails: individual freedom, political equality, economic and educational reform, and, significantly, accountability. Democratization was hailed as the instrument to obliterate the hotbeds of terrorism throughout the region and in the periphery.

The Bush administration has acknowledged repeatedly that the process is arduous and cumbersome, with uncertain prospects for success. Broad-based grassroots resistance to the transformation pervades the Arab Islamic world, and even the Europeans – would-be natural partners to such an initiative – have been reluctant to enlist in the effort. Furthermore, the Bush administration has uncovered other difficulties in trying to translate its vision into reality. First, it has been beset by the need to determine how much pressure should be leveled against existing regimes. The Americans are concerned that excessive pressure will undermine these regimes. However, implementing merely symbolic democratic measures such as elections, without first addressing the deeper, rooted problems within the Arab world, may well result in the rise of radical Islamic anti-American regimes along the lines of the Iranian model. The fact is that the Islamic groups in these countries are better organized than other groups and enjoy a relatively positive popular image. A simple "one man–one vote" system that is ungrounded in democratic civic institutions could easily propel authoritarian elements into seats of power. Second, it is clear to the Bush administration that change must come from within. The United States can assist and energize, but it cannot be perceived as imposing the change, and certainly not as circumventing or undercutting Islam. Yet the internal forces supporting reform in the direction advocated by the Bush administration tend to be weak, typically consisting of groups of intellectuals with no political base and little mass appeal.

Despite the difficulties, there have been glimmers of progress. The current of change emanating from Washington over the past few years was undoubtedly a critical ingredient fueling the substantial wave of protest that has recently swept the Arab world. The overall impression is that for the first time in many years it is possible to discern, even if only to a limited degree, the voice of what can be called "the public," along with initial adherence to international standards and norms.

In Lebanon, mass protests and the supporting role of the United States and Europe (especially France, which in itself is noteworthy) led to Syria's hasty decision to withdraw its forces from Lebanon after an occupation of nearly thirty years. This development capped the steadily growing Lebanese opposition to Syria's exploitative rule in the country. This in turn had accelerated against the background of the IDF's withdrawal from southern Lebanon in 2000, which obviated the pretext for any Syrian presence and undercut the radical message transmitted by Syria and Hizbollah. Damascus, radical out of convenience and not necessarily out of conviction, used the radical agenda to secure and empower its interests in Lebanon. Its insistence on extending Emile Lahoud's term as president of Lebanon created a snowball effect that began with UN Security Council resolution 1559 and ended with the assassination of Rafiq al-Hariri, which stirred public outrage and the demand that Syrian troops withdraw from the country. To be sure, even greatly weakened by the absence of its military presence in Lebanon and after elections and the establishment of a new government, Syria has remained the dominant power in the country. What has changed significantly is the standing of Hizbollah, which, by emphasizing its Lebanese identity and by increasing its involvement in the country's political system, has fought to retain its right to exist as an armed organization. Hizbollah contends that its military arm provides Lebanon with deterrence capability and creates a strategic balance with Israel.

In any event, Syria has been weakened and isolated – a situation caused in part by President Bashar al-Asad and his lack of political savvy. In the five years of his presidency, he has fulfilled few of the expectations that were created upon his ascendance into power. Asad, who apparently still possesses the loyalty of the top Alawi leadership, has barely preserved some of the achievements secured during his father's tenure, yet he has not been able to extricate himself from their constraints. For example, he has continued to support Hizbollah and to encourage Palestinian attacks against Israel, but in contrast to his father he has not refrained from encroaching on American interests. Syrian weakness has led Asad to signal willingness to begin negotiations with Israel, an initiative received suspiciously by Israel as no more than a tactical step to improve Syria's image. Israel also doubts whether a weak Bashar will be able to be more flexible during negotiations than his father.

In a related development, the long-term stability of a number of regimes throughout the region came into question during 2004–5 for the first time in many years. The strength of the Saudi regime continues to ebb slowly. As Paul Rivlin notes in his essay, rising oil prices have eased the Saudi regime's problems somewhat and have expanded its internal room to maneuver. However, the Saudis have now been forced to address domestic terrorism, which appears to have proponents within the Saudi administration itself. The degree to which the regime is currently succeeding in its battle against terrorism is still unclear.

Egypt, and to lesser degree Saudi Arabia, has responded to American pressure and has agreed to make a number of cosmetic changes associated with democratization. It hopes this will spare it any serious makeovers to neutralize the problems identified by the Bush administration. Mubarak, who even if he earns a fifth presidential term in September 2005 may be approaching the end of his time at Egypt's helm, has also been challenged by increasing terrorism, strained relations with the civilian population, and general public dissatisfaction.

On the other hand, the year has witnessed a tangible improvement in Egyptian–Israeli relations, coinciding with the reduced violence between Israel and the Palestinians and Israel's disengagement initiative. After a long hiatus, a new Egyptian ambassador to Israel was appointed. The two countries signed a strategic deal whereby Egypt will supply Israel with a significant portion of its natural gas needs. Egypt has also increased its role as mediator between Israel and the Palestinians and has expressed conditional willingness to assume responsibility for barring smuggling from its territory into the territory of the Palestinian Authority, primarily along the Philadelphi route.

Above all, however, it is Iraq that is viewed by the Bush administration and by regional actors as a test case and a yardstick for assessing the ability of the United States to translate the vision of political transformation into practical language.

Noted progress in building the new regime and its institutions is evident. Important strides in restoring life to normal have been made in economic reconstruction and the resurrection of the education and health systems. Furthermore, the January 2005 elections, which were the first democratic elections ever held in Iraq, can certainly be heralded as an achievement, despite limited Sunni participation. Nonetheless, as Ephraim Kam's chapter argues, more than two years after the toppling of Saddam Hussein's regime, and in light of the heavy human, economic, and political costs of the war and its aftermath, the success–failure balance sheet of American actions is mixed at best.

Iraq is currently plagued by instability, intense violence, terrorism, and internal conflicts within and among the various ethnic groups that comprise its population. Members of most of Iraq's ethnic groups have assumed some sort of role in this self-perpetuating cycle of violence and terrorism. The largest role, however, has been played by Sunnis loyal to the previous regime, who have not fully internalized their loss of power and who have not been willing to accept the Shiites as the leading force in the new regime. Also active in the violence have been Islamic elements, including Shiites. Some, originally from outside the country, have enlisted to fight a "holy war," motivated by nationalist and primarily religious ideas of hatred of the West (especially the United States) and resistance to foreign occupation. Some of these activists have regarded the war with the United States as a confrontation with Israel as well, and some groups have hoped that it might be possible to take advantage of the chaotic state of affairs in order to establish an Islamic regime.

Indeed, terrorism, which for the most part has been the work of radical Islamic elements, has continued to pose a challenge that is trans-border and trans-national in nature. It appears that the blow suffered by al-Qaeda following the attacks of September 11 did not entirely quash its ability and the capacity of groups like it to execute showcase attacks both in Western and Muslim countries. Despite improved inter-state coordination in the war on terrorism, Spain, Turkey, Saudi Arabia, Egypt, and Great Britain were victims of severe attacks in 2004–5, and Russia suffered terrorist strikes by the Chechen underground. Hatred of the West drives the actions of the terrorists, and the war in Iraq and the presence of foreign forces in the country have served as pretexts for legitimizing attacks on "the Christian–Jewish alliance," which also includes local collaborating regimes.

Unfortunately, as Yoram Schweitzer points out in his essay on international terrorism, no change appears to be on the horizon in this sphere. The concern is that Iraq will serve as fertile ground for terrorist training and practical experience – like Afghanistan during the 1980s – and that from there terrorists will export their expertise to other countries. The threat of another mega-attack by conventional or non-conventional means requires increased cooperation among countries in the international war on terrorism. Terrorist attacks and the prevalent public sense of vulnerability has sparked increased xenophobia, most notably in Europe, and led to a series of emergency measures unprecedented in Western democracies.

The establishment of a fundamentally new regime and the suppression of terrorism (with the latter serving as a prerequisite for the former) are the key criteria for success of American activities in Iraq. The Americans are aware of the far-reaching implications of failure in Iraq. Despite the heavy costs of their involvement, the Bush administration, enjoying the breathing room provided by a second term in office, is resolved to maintain its presence and advance its recipe for reconstruction. There is not enough domestic pressure to compel the withdrawal of forces from Iraq any time in the near future – that is, not before America achieves some degree of stabilization. However, heavy losses and a feeling that there is no way out are eroding what has until now been considerable American public support for ongoing activity in Iraq. Calls for the government to devise an exit strategy have intensified.

Iran has been following developments in Iraq closely. For the past few years, the regime of the ayatollahs has exhibited nascent signs of decline. However, the more conservative elements in the regime gained strength in 2004–5, as reflected in the parliamentary and presidential elections. Still, the winds blowing out of Tehran are becoming more nationalistic and less religious. As for its nuclear program, in both word and deed the Iranians seem determined to acquire military nuclear capability. During the past year, they focused on diplomatic maneuvering in the hope of gaining time to progress on the project, and they have not acceded to European suggestions to suspend nuclear development in exchange for economic incentives. They have sought a way to divide the Europeans from the Bush administration, parties that until now have acted in close coordination. Iran prefers that the nuclear issue not reach the Security Council, and hopes that nuclear progress will deter the United States from exerting military pressure, an option that the Americans have not ruled out. There are virtually no disagreements between the involved parties regarding the dramatic and dangerous implications for regional stability stemming from Iran's potential for achieving military nuclear capability. But some voices doubt the possibility of stopping Iran without the use of military force. Like the United States, Israel has kept a low profile on the issue and, at least for the moment, prefers a European-brokered diplomatic solution to the crisis.

The nuclear activity of the Iranians, as well as that of the North Koreans, has underscored the problems of the Nuclear Non-Proliferation Treaty (NPT). Thus, along with the challenge of Iran comes a challenge to the non-proliferation regime. As Emily Landau observes in her chapter on Iran and non-proliferation challenges, ideas have been put forward to strengthen the treaty, but the May 2005 NPT Review Conference ended in failure due to an inability of states to agree upon non-proliferation priorities.

II

While much of the Arab and Muslim Middle East has been challenged to reposition itself according to the political goals of the Bush administration, Israel has been challenged by its own new political reality. In 2004–5, Israeli politics and society revolved around the flurry of unilateral disengagement.

In early June 2004 the Israeli government approved an amended version of the plan for disengagement from the Gaza Strip and four settlements in northern Samaria, first announced by Prime Minister Ariel Sharon in December 2003 and presented to the Knesset three months later. The government's decision and its approval by the Knesset were attained only after an intensive political struggle and changes in the makeup of the government coalition. Sharon lost his parliamentary majority and has had to deal with an ongoing revolt in his own political party. It was only with Labor's entry into the coalition – which, by its nature, is merely a temporary marriage of convenience – that Sharon was able to overcome the severe intra- and extra-parliamentary opposition and translate the initiative into reality.

The disengagement plan, which marks the first time that the State of Israel has decided to evacuate settlements in the Land of Israel, embodies major political, economic, and social significance. On a strategic level, it represents a dilution of the ethos that Sharon himself, "the father of the settlement movement," nurtured regarding the settlements' strategic contribution to Israel's future borders. It unequivocally slashes Greater Land of Israel aspirations, at least in their most comprehensive parameters. At the same time, the disengagement reflects the Israeli public's increasing awareness that the growing demographic threat of the Israeli-occupied territories mandates that if Israel is to remain a Jewish democratic state, its control over the Palestinians must yield to a two-state solution by partitioning the land.

Equally if not more important than the disengagement plan in enhancing the new understanding on the need for partition is the separation fence. The fence, a proven and recognized contribution to the war on terror, in effect demarcates potential borders between Israel and the future Palestinian state. Portions of the fence bear some resemblance to the Green Line, while other sections, some still in dispute, veer eastward to enclose settlements within its perimeter. The future of the West Bank settlements and major settlement blocs, which have earned much greater Israeli popular support than the Gaza settlement enterprise ever did, is a major political challenge to both Israelis and Palestinians alike. But however the issue is ultimately decided, the intellectual reality of separation is increasingly engraved in Israeli consciousness through the tangible reality of the separation fence.

Support and opposition for these two major initiatives have regularly crossed partisan lines. While much of the right has generally rejected the disengagement, some of the opponents were joined by left-wing compatriots in protesting the unilateral nature of the plan. In other words, the withdrawal from Gaza was less problematic to some than was the lack of an agreement and the absence of anything in exchange for the evacuation. For its part, the separation fence has had various political patrons since its initial emergence as a child of the left, which was eager to effect a separation from the Palestinian population, and its later championship by the right as a valuable deterrent against terrorism. Construction of the fence lagged last year for a number of reasons, including American opposition to creating facts on the ground via the fence's demarcation. Opposition from the left wing that led to the Supreme Court's intervention and legal appeals over the fence's route and its harsh effect on many Palestinians communities caused additional delays in construction. But from any angle, it is clear from both the disengagement and the fence that the reality of separation is under both physical and ideological construction.

Echoing the political divide have been other rifts. The heavy representation of the religious Zionist community among the opponents of the disengagement has suggested to some that the disengagement has evolved into a religious–secular issue, even though the opposition to the disengagement among the general public far exceeded the percentage of the religious population. There have been those who argued that the opposition of the religious camp to the government decision created a dichotomy that reflects the historical clash between "monarchy" and "priesthood" and embodies the fierce debate over the essence of the State of Israel. The secular public grew alarmed by the legitimacy that members of the religious camp have conferred upon exhortations to the settlers to defy the state's institutions. Some within the religious sector have seen the disengagement as symbolic of their alienation from the state, and there has been extensive soul-searching within the greater religious community as to how to assimilate the initiative and the social divisions expressed in its wake, both among themselves and with the secular majority.

Also poignant has been the issue of the Israel Defense Forces (IDF), which has long been hailed as a pillar of national consensus. The fact the IDF, by virtue of its operational involvement in the disengagement, has been forced into a bitter social and political debate, has punctured a basic unifying link in Israeli society. The sensitivity of the issue has been aggravated by those refusing certain tasks in the IDF. Whereas once the headlines were dominated by those who felt it immoral to serve in the territories and refused to support an occupying army, this year's stage has focused on those among the right who have called for widespread refusal to evacuate settlements – encouraged at times by religious and political leaders. Some of the inter-sector rifts have been deemed sufficiently disturbing to spawn attempts at social dialogue and mutual understanding, while some have challenged the very possibility of finding common ground. The stakes are high not only for the immediate disengagement agenda and the social reconstruction that must follow, but for future high-intensity issues as well, chief among them the future of the settlements in Judea and Samaria.

On the international scene, the disengagement plan has granted Israel a small measure of comfort, particularly in its relations with Europe, frequently characterized by discord. European states have begun to exhibit a greater understanding of Israeli policy, and have been willing to consider, more positively than in the past, European participation in promoting an arrangement with the Palestinians. For its part, the United States has expressed support for the disengagement, provided that it help jump-start the roadmap and not signal an abandonment of the Quartet plan, which has likewise been accepted by Israel and the Palestinians.

As part of the close ties between Israel and the United States and because of its own entrenchment in the war on terror, the Bush administration essentially gave Israel a free hand in dealing with Palestinian terror. However, diplomatic friction that arose between the two countries this year illustrates the sensitive nature of the overall alliance. The dispute centered around the familiar issue of Israeli military exports, this time once again regarding the sale of controversial military equipment to China, and prompted the United States to take punitive steps in its security relations with Israel. Yet beyond the issue of military exports lies the broader issue of "Israeli credibility," which will no doubt surface following the disengagement *vis-à-vis* Israel's promise to freeze settlement construction and dismantle illegal outposts in the West Bank.

As Shlomo Brom observes in his chapter on the Israeli–Palestinian conflict, highly significant for the potential of the disengagement plan are the disintegration of the militarized intifada, especially in Judea and Samaria, and the emergence of a new Palestinian leadership following the death of Yasir Arafat in November 2004.

Years of terror and high casualties spawned a vicious cycle of response and counter-response that impacted heavily on the Israeli and Palestinian societies, both of which have grown tired of the price exacted from them. Israel emerged victorious from the military confrontation with the Palestinians, but became increasingly cognizant of the limitations of its power, namely, that force was no alternative to a political arrangement. The intifada proved to the Palestinians that although Israeli society had been unwilling to pay a (relatively low) price for remaining in Lebanon, it was willing to pay a very high price not to yield to Palestinian demands. At the same time, while Israel was able to reduce the violence to a more "tolerable level," it could not diffuse the motivation underlying the intifada. Hence the growing need for a political initiative.

Arafat died days after the Israeli Knesset approved the unilateral disengagement plan. Founder and leader of the Palestinian national movement and long-time advocate of the armed struggle, Arafat was succeeded by Mahmoud Abbas (Abu Mazen), who was elected in legitimate democratic elections. Abbas has emphasized the futility of violence and the value of negotiations with Israel. Yet while the Palestinian population, like the Israeli population, is tired of the armed intifada, Abbas has far from secured his political base in the Palestinian system. He faces a strong challenge both from Fatah rivals and Hamas, and fear of Hamas' popularity prompted him to postpone the long-awaited parliamentary elections. Abbas's intention and ability to move against Hamas, Islamic Jihad, and terrorism's infrastructure, required of him by the roadmap, are highly doubtful, especially while Hamas continues to garner support and has even begun to consider limited participation in the Palestinian Authority, while not reneging on its destructive objectives *vis-à-vis* Israel

The many questions raised by the disengagement remain and await resolution. The two sides' respective needs from the disengagement diverge so radically that they are liable to foster a crisis of unmet expectations. Each leader faces enormous domestic political challenges and will find it well-nigh impossible to satisfy his counterpart's demands. Abu Mazen will probably find it difficult to act forcefully enough in the war on terror, and Sharon will also have a hard time fulfilling Abu Mazen's demands for additional withdrawals. This is regardless of whether elections in Israel are advanced because of the dissolution of the "disengagement coalition" or are held in November 2006, the scheduled date. It is thus doubtful that the coming year will witness additional major steps toward an Israeli–Palestinian arrangement. In that case, the main problem will be how to keep the crisis from spiraling out of control and how to prevent a revival of the militarized struggle.

III

This is the background to *The Middle East Strategic Balance 2004–2005*, which launches the third decade of publication of the Jaffee Center for Strategic Studies annual. The primary goal of this book is to review some of the important regional developments and issues that impact directly on Israel's strategic options.

Part I of the present volume opens with Mark Heller's overview of the winds of change blowing in the Middle East, fanned by the United States' drive for greater democratization in the Arab and Muslim worlds. The following three chapters zoom in on more specific arenas. Chapter 2 is Ephraim Kam's analysis of the ongoing crisis in Iraq, a volatile situation that impacts heavily on the domestic, regional, and international scenes. Chapter 3 moves from the locus of the United States' main effort in the Middle East, Iraq, to another central concern: the Israeli–Palestinian conflict. Shlomo Brom analyzes the glimpses of new reality emerging in the conflict, shaped primarily by the end of the militarized intifada, the emergence of a new Palestinian leadership, and Israel's efforts, through the disengagement plan and the security fence, to separate from the Palestinians. In chapter 4, Emily Landau examines ongoing international efforts to preempt the possibility of a nuclear Iran, and then extends the discussion to consider broader arms control efforts that might help the world deal with Iran in the event that it indeed becomes a nuclear weapons state.

Each of the two chapters that follow explores a particular subject intrinsic to Middle East developments. In chapter 5, Yoram Schweitzer examines recent trends in international terrorism, assessing the ongoing ability of the international jihad movement to wage successful attacks, both in the Middle East and in Europe. The final chapter of this section, written by Paul Rivlin, explores the repercussions of higher oil prices on Middle East states. The section concludes with a chronology of major events from July 2004 to June 2005.

Part II of this volume, compiled by Yiftah Shapir with the assistance of Tamir Magal and Avi Mor, offers a review of the region's military forces through brief assessments of each country's military resources. Emphasized here are major changes to the orders of battle and key components of the individual force structures. More detailed data on the inventories of military forces appears on the Jaffee Center website and is updated on a continuous basis. Support for the research and updating the data online is made possible through proceeds from the Dr. I. B. Burnett Research Fund for Quantitative Analysis of the Arab–Israeli Conflict.

I would like to express my gratitude to members of the Jaffee Center research staff who made the publication of this volume possible; to Moshe Grundman, Director of Publications at JCSS, who coordinated all aspects to the publication of this volume; and to Judith Rosen, Editor at the Jaffee Center.

ZVI SHTAUBER
Head of JCSS
Tel Aviv, July 2005

PART ▶▶▶

I

Middle East
Strategic Assessment

CHAPTER ➤➤➤

1

The Middle East:
Still on the Brink of a New Era

MARK A. HELLER

In 2004–5, the political order in the Middle East was shaken to a greater degree than at any time since the Islamic Revolution in Iran. No regimes were overthrown, and with the partial exception of Lebanon, none were even seriously threatened. At the same time, a wave of unrest and protest did unnerve the established order, put status quo forces on the defensive, and impart a new immediacy to the feeling that the region was ripe for far-reaching change. And perhaps for the very first time in modern history, the anti-establishment forces included not just radical elements (national-socialist or Islamist) but also significant pro-democracy components.

The United States was eager to take credit for this effervescence. After all, President Bush had explicitly renounced the traditional American preference for stability and committed himself instead to the cause of freedom, which implicitly meant subverting and/or alienating authoritarian governments, even those otherwise deemed friendly to the United States. In fact, however, much of the demand for change came from domestic forces, although it is also true that those forces were at least partly encouraged and empowered by outside inspiration. The American-led destruction of Saddam Hussein's "Republic of Fear," the most dramatic of catalysts, was joined by other drives to open expression of the discontent that had been suppressed or repressed for many years.

There is a certain irony in the fact that the stimulus for change was provided by external engagement in the region, because when the sense that the Middle East was on the brink of a new era first emerged after the end of the Cold War, it was prompted then by the opposite expectation – that outsiders, no longer interested in the region as a strategic asset, would leave it to its own devices for the first time since Napoleon Bonaparte tried to export the French Revolution to Egypt in 1798. Analyses of that sort did not imply any specific conclusions about the direction the region would ultimately take, but they did suggest that Middle Easterners would henceforth be both free to determine their own political, social, and economic destinies without foreign interference and responsible for whatever consequences emerged from their choices.

Subsequent developments have tended to indicate that such prognoses were, if not misplaced, then at least premature. Rather than ushering in an era of autonomy and self-direction for the Middle East, the end of the Cold War marked the beginning of more, and in some ways, more intrusive extra-regional influence and presence, all of which have contributed to much greater destabilization than great power political rivalry ever did.

➤ EXTERNAL PENETRATION AND PRESSURE FOR CHANGE

One manifestation of external penetration has been the relentless force of post-Cold War globalization, which has prompted massive cultural, economic, and technological infiltration of the region. The years since 1990 have witnessed not only the departure of the Soviet Union from the Middle East, but also the arrival of satellite television, cellular phones, and the internet. For the most part, status quo regimes and conservative social forces have failed either to resist this penetration or to adapt successfully to it, and the result has been persistent systemic dysfunction, amply documented by a series of human development reports by the United Nations Development Program.

The second stimulus of change has been purposeful penetration – the conscious resolve by the West to intervene directly in response to the export of this dysfunction. In some cases, that intervention has had an economic–administrative character (under the guise of development assistance/partnership programs); in others, it has been of a blatantly political–military nature. Economic–administrative intervention was more typical of European efforts, particularly under the umbrella of the Euro-Mediterranean Partnership, to encourage economic and political–social reforms on the southern and eastern rims of the Mediterranean in order to reduce the spillover of the area's dysfunction into Europe, in the form of terrorism, crime, and especially illegal immigration. Political–military intervention was identified more with American policies, even though the United States also pursued developmental aid efforts of its own and elicited various degrees of European support for political–military activism.

The first dramatic post-Cold War example of political–military intervention was the American response to the Iraqi invasion of Kuwait in 1990. Because the campaign involved the ouster of Iraq from Kuwait but not the overthrow of Saddam Hussein, it was followed by the permanent deployment of a large American military force in Saudi Arabia to protect the kingdom against future Iraqi threats. But since the American "occupation" of the Land of the Two Holy Places became the primary grievance cited by Osama bin Laden in his 1998 declaration of jihad against the West, the liberation of Kuwait – a massive intervention aimed at preserving the regional status quo – actually triggered a chain of events that eventually led to 9/11. In turn, 9/11 generated the conviction that terrorism was the quintessential export of dysfunction. If so, the logic held, the "root cause" of the terrorist threat to American security was the deformation of political society in the Arab/Muslim world, a deformation that could only be rectified by the aggressive promotion of wholesale transformation under the rubric of "democratization."

The material consequence of this conviction was that in 2004–5, Western military forces were more widely deployed in the broader Middle East than when the Cold War ended fifteen years earlier. NATO troops were still operating in Afghanistan following the elimination of the Taliban regime in 2001 in order to help protect what was at least an embryonic democracy led by Hamid Karzai, the appointed interim leader whose presidency was confirmed by popular election in October 2004. And American and allied Western forces have maintained a huge presence in Iraq in an effort to restore security after felling the Saddam Hussein regime in 2003, an operation ostensibly driven by Saddam's alleged defiance of Security Council resolutions concerning weapons of mass destruction but in truth almost certainly motivated – and rationalized *ex post facto* – by the desire to eliminate the most blatant case of political deformation in the region.

Another and perhaps ultimately even more significant consequence was the sense of political ferment throughout the region that Western intervention of all sorts has helped to stimulate. More than the aftermath of regime change in Afghanistan, this ferment was stimulated by the events in Iraq. It would be an oversimplification, of course, to claim that Western intrusion, reactive or proactive, was the sole catalyst to this ferment. On the contrary, internal pressures, coupled with economic imperatives, had produced some cautious experiments in political and/or economic liberalization in the late 1990s in Iran, Jordan, Morocco, and some of the Gulf sheikhdoms. Even in Syria, there was a brief "Damascus spring" immediately after the death of Hafez al-Asad in 2000, during which freer expression of opinion was tolerated by the authorities. But in all these cases, the regimes held the reins that drove the degree and pace of change, and in most of them, concern about loss of control soon prompted the regimes to try to retrench, which they did with impressive ease.

After September 11, however, the increasingly muscular American posture generally helped produce a more sustained dynamic of change. This did not mean that the United States or other Western powers were imposing their own values or political systems on other countries, as critics often charged. But it did mean that they were energizing domestic proponents of change – even as the latter strove to avoid being tainted as "agents" of foreign powers – by putting status quo regimes on the defensive and, in the most extreme cases, eliminating them as the obstacles to change. As a result, the period since 9/11, and especially since the American-led ouster of Saddam Hussein, has witnessed a far more vigorous indigenous debate than ever before about the shortcomings of the status quo and the need to pursue far-reaching alternatives, as well as the adoption of noteworthy measures – even if many were only cosmetic – intended to demonstrate recognition by the regimes of the need for change. In 2004–5, these dynamics were evident across the entire region, although they manifested themselves in different ways.

➤ REGIONAL RAMIFICATIONS

➤ Iraq

The most direct effect, of course, was in Iraq itself. The American-led invasion unleashed a maelstrom of pent-up ethnic, religious, and political demands hitherto suppressed by the brutally effective machinery of central control. It also provoked a persistent campaign of violence by a mix of elements – former Ba'athists, other disenfranchised Sunnis, and foreign Islamists – against the foreign presence, its local collaborators, and rival sects or communities. Moreover, concern about Iraq's ultimate direction stimulated competitive intervention by many of Iraq's neighbors in hopes of influencing the evolving political system, preempting influence by other states, or simply ensuring the perpetuation of ongoing chaos, either to compel the United States to leave in defeat or, alternatively, to keep it bogged down permanently so that it could not turn its attention elsewhere. Indeed, the list of things that could potentially go wrong and stay wrong as a result of regime destruction in Iraq was virtually endless, and long-time observers of Middle Eastern politics could justifiably believe that many of them would.

But along with these worrisome implications, the elimination of Saddam's regime opened up space for non-violent self-expression and demands for political participation by a whole host of previously disenfranchised individuals and groups, i.e., for the critical first steps in real democratization. It is ironic that this space was created by belligerent foreign intervention and was tested in parliamentary elections held in January 2005 under what essentially remained foreign occupation. Nevertheless, and notwithstanding determined efforts by opponents of democratization to disrupt the election, including threats that left little to the imagination, millions of Iraqis defiantly turned out to register their approval of the process by casting a ballot.

The Iraqi election, of course, hardly signified an irreversible step in a smooth and inevitable evolution toward full-fledged democracy. The voting could only proceed under the watchful eye of some 300,000 foreign and Iraqi security forces, and voting patterns confirmed widespread alienation among Sunnis, the existence of political currents among majority Shiites that included strong Islamist sentiments not necessarily compatible with values of democracy and pluralism, and the unswerving insistence of Kurdish representatives on a degree of autonomy that might not be accommodated in negotiations for a constitution scheduled for later in 2005. Many of these complications were evident in post-election coalition negotiations, which needed several months to produce a mutually acceptable government.

Still, the regional implications of the popular reaction to this Iraqi experiment in democracy were clear. It encouraged others in the Arab Middle East to demand that their own rulers give them the same opportunity given to Iraqis by foreigners, and it put ruling regimes pursuing policies and/or systems of governance similar to those of Saddam on notice that Western renunciations of "benign neglect" were not mere lip-service, and that even if repetition of an Iraq-style invasion was improbable, refusal to change would have some sort of undesirable consequences.

➤ Libya

The most straightforward reaction came from Libya, which essentially concluded that it had to abandon its weapons of mass destruction development programs and renounce support of terrorism in return for normalization of relations with Western governments, including the United States. American analysts generally attributed this policy shift to the demonstration effect of American action in Iraq; Muammar Qaddafi, it was argued, resolved to appease the United States in order to avoid the fate of Saddam Hussein. European analysts tended to assign greater weight to the cumulative impact of UN-approved economic and diplomatic sanctions, as if to suggest that the fact that this impact achieved critical mass only after the ouster of Saddam was sheer coincidence. In any case, Qaddafi was primarily troubled by external rather than domestic pressure, though he did face an Islamist challenge at home, and he secured Western acquiescence in a fairly simple bargain: policy change in return for easing of pressure for regime change.

➤ Syria and Lebanon

The regime in Syria may well have wanted a similar deal. While not a charter member of the "axis of evil," it shared enough characteristics in common with Saddam's regime – authoritarian rule by a minority ostensibly based on the Ba'ath Party but in reality relying on the security services to repress opposition; pursuit of anti-American policies; suspicions of covert development of weapons of mass destruction; and support of terrorism – to render it a deformation only slightly less skewed in American eyes than was Iraq and to exacerbate fears that Syria was next on the American target list. As a result, the Syrians became more forthcoming on terrorism-related intelligence cooperation with the Americans, especially about threats connected to al-Qaeda. The problem, however, was that they could not truly accommodate American demands on other issues without incurring seemingly unacceptable risks to regime survival. The most critical of these issues had nothing to do with the peace process with Israel. Instead, they related to Iraq and Lebanon. In response to American complaints that most of the foreign fighters and much of the money fueling the insurgency in Iraq were routed through Syria, the regime argued that it was trying to control the border but simply lacked the necessary capacity. But if Syria's ability to close the border was a matter of some dispute, so too was Syria's interest in doing so. After all, as long as the United States remained bogged down in an inconclusive struggle in Iraq, it would not have the resources to implement the program that at least some Syrians believed was in store for Syria. In any case, Syrian action in this regard was insufficient to eliminate it as a point of contention.

The same is true with respect to involvement in Lebanon. There, crude Syrian intervention secured a constitutional amendment in mid-2004 permitting the extension of the term of office of President Emile Lahoud, widely seen as a Syrian cat's-paw. In the face of growing Lebanese opposition to continued Syrian control of Lebanon, some such measure may have been necessary to preserve what was perceived as a critical strategic asset (and a resource to permit the continued enrichment of important power bases in the Syrian security establishment), especially if Syria was considering some force drawdown in order to appease the United States. But this one was carried out with such heavy-handedness that it backfired and precipitated a greater wave of anti-Syrian protests. Even more astonishing was the unusual spectacle of French joint sponsorship with the United States of another Security Council resolution demanding the complete withdrawal of Syrian forces from Lebanon, this time coupled with a demand for the dismantling of all armed militias, i.e., of Hizbollah.

Bashar al-Asad's actions in this regard strengthened suspicions that he lacked the basic qualities of his father needed to rule effectively. If Bashar's faults were immaturity and inexperience, they might be overcome in time; if the flaw was simple-mindedness, it could be permanent, and mortally dangerous to the regime. The subsequent assassination of former prime minister Rafiq al-Hariri in February 2005 appeared to confirm these suspicions. Whether or not Hariri was killed at direct Syrian behest, it was widely believed in Lebanon and elsewhere that Syria was behind the assassination, notwithstanding adamant denials by Syrian spokesmen and the virtual impossibility of proving the charge, if only because most of the physical and forensic evidence disappeared in the car bomb blast that killed him. After all, Hariri had resigned in protest against the extension of Lahoud's term and had reportedly used his widespread international contacts to mobilize support for the Security Council resolution that followed. The immediate result of the assassination was a huge outpouring of public outrage that crossed most of the communitarian boundaries, with only the Shiites, mobilized by Hizbollah, rallying to Syria's defense. Opponents of Syria's presence in Lebanon and of the Lebanese political leadership and security forces that cooperated with it turned out in massive numbers to launch what they called an "independence intifada" or, inspired by the peaceful revolt that toppled incumbent rulers in Ukraine shortly before, a "Velvet Revolution" (though some cynics, more impressed by the social character and dress of many demonstrators, preferred to described it as the "Gucci Revolution"). But whatever term is used to describe the protesters' actions, they did help bring about the ouster of Lahoud's pro-Syrian prime minister, the resignation of some Lebanese security service commanders, and, most significantly, the eventual withdrawal of all Syrian military forces at the end of April.

The Syrian military withdrawal will not necessarily eliminate all Syrian involvement in Lebanese politics, and persistent reports that Syrian intelligence agents continued to operate in Lebanon reflected the belief that Syria would continue to search for other avenues of influence. Nevertheless, direct military presence was the most potent lever, and the liquidation of that presence will have important implications for the configuration of the Lebanese political system following the elections in late May–early June that gave the opposition parties a parliamentary majority. The end of direct Syrian control is particularly relevant to the character and role of Hizbollah. Because the Shiites do not comprise a majority, Hizbollah cannot realistically aspire to dominate all of Lebanon. But even if it only seeks to champion Shiite interests, it will be obliged to engage in coalition politics with other Lebanese factions, and that may force it to accommodate demands by other Lebanese to disarm its militia and abnegate its traditional identity as a resistance movement.

However Hizbollah deals with this potential dilemma, the consequences of the Syrian withdrawal may also extend to Syria itself. By the time the withdrawal was finally implemented, it was no longer sufficient – if it ever had been – to divert pressures for domestic change, either from abroad or from within. It is debatable whether significant elements within the Syrian establishment view the withdrawal as a capitulation for which Bashar should be challenged, or whether the loss of economic assets, such as access to the Lebanese labor market for surplus Syrian manpower, will be severe enough to have political ramifications.

In any event, it is already clear that embryonic civil society elements were affected by events in Lebanon and even further away in the international arena. A number of Syrian intellectuals, for example, were emboldened to give unprecedented public expression to their identification with the Lebanese opposition and the values it ostensibly upheld. Their actions were preceded by even more assertive protests by Syrian Kurds at the deprivation of citizenship rights that they endured for years. Taken together, these developments suggest that the regime will continue to be exposed to further pressure for change in both foreign policy and governance, from both foreign and domestic sources. Asad's stammering and highly equivocal endorsement of liberalization and reform at the Ba'ath party conference in June was the first indication following the withdrawal from Lebanon that the regime has no clear and coherent strategy to deal with this pressure.

➤ The Palestinian Authority

In many important respects, the Palestinian Authority has resembled Syria in recent years in the sense that while the PA, too, was not classified as part of the "axis of evil," it too has confronted serious American and even growing European pressure to change both policy and governance. The demands with respect to policy have had less to do with declared Palestinian objectives *vis-à-vis* Israel than with the means by which they were being pursued; since 9/11, there has been dramatically lower Western tolerance for terrorism no matter what the pretext. The demands with respect to governance have had to do with transparency and accountability in the operations of the Palestinian Authority, along with political openness and the rule of law. These issues were at the center of Bush's decision in 2002 effectively to suspend any American involvement in the Palestinian–Israeli conflict and to boycott the PA until the Palestinian leadership was replaced by one untainted by repression, corruption, or terrorism. But as in the case of Syria (and unlike that of Iraq), the major obstacle to change here was eliminated, not by proactive American measures but by natural causes: the death of the leader.

Still, there were also major differences between these two cases. The Palestinian Authority was less crudely repressive and less willing or able to isolate Palestinian society from the outside world, meaning that Palestinians were more aware of the collective/national as well as individual costs of the failure to change. As a result, the indigenous Palestinian demand for domestic reform and democratization was far more widespread or at least more audible than in Syria; in many respects it actually preceded American and other foreign interest in such matters. As a result, the death of Yasir Arafat in November 2004, unlike the death of Hafez al-Asad more than four years before, removed an obstacle to pent-up domestic forces favoring the very changes demanded by external actors, even as Palestinian reformers went to great lengths to dispel accusations that they were effectively collaborating with the United States or, worse, with Israel.

These forces found some expression in the January 2005 election that installed Mahmoud Abbas (Abu Mazen) as Arafat's successor. Like the election in Iraq, it took place under the shadow of occupation but also under the watchful eye of large numbers of international observers who confirmed its free and fair nature. The relatively large turnout, despite the boycott by Hamas, reflected general endorsement both of Abbas's specific policy proposals – especially the end of the "armed intifada" and the rationalization of the PA's government machinery – and of the broader principle of determining political leadership and succession by democratic means. In that sense, the death of Arafat can be seen as having already signaled a paradigm shift, one that was seemingly reinforced by municipal elections in the spring of 2005 that were marked by some charges of irregularities but generally conformed to prevailing standards of democratic practice. However, the durability of that shift will be further tested in legislative elections in which Hamas will participate, in the formation and practices of post-election governments, and – on policy issues – in the Palestinian approach to Palestinian–Israeli relations after the Israeli disengagement from Gaza and the northern West Bank.

Furthermore, the Palestinian experience, however it evolves, does not and cannot signify much for the rest of the region, superficial similarities with Syria notwithstanding. It may be true that aspiring democratizers elsewhere noted with some envy the fact Palestinians had the opportunity even under Israeli occupation to vote freely in a real election that they themselves were denied. But analogies are constrained by the fact that the protracted conflict with Israel has pervaded every aspect of Palestinian public life and led to social, political, economic, and psychic interaction with the outside world to a degree rarely found elsewhere in the Middle East.

➤ Saudi Arabia

More typical are the experiences of Saudi Arabia and Egypt, where foreign policies already more or less conform to American preferences and policy change therefore cannot be invoked as a means to forestall pressure for real changes in governance. These two countries are American allies of long standing but they are also the home of the September 11 hijackers. That simple fact summarizes the failure of the traditional American policy of support for cooperative but authoritarian regimes. George W. Bush began to renounce and denounce the policy of purchasing stability at the price of freedom as far back as 2002, and he repeated this position in his 2005 State of the Union address. But this time, for the first time, he specifically singled out, though with non-confrontational rhetoric, Saudi Arabia and Egypt as countries that should be moving in the direction of greater democracy.

In fact, both countries did undertake experiments in democratization, though with less clarity and consistency than critics would have liked. In Saudi Arabia, the ruling family felt the need to respond to stirrings of liberal dissent (e.g., demands for greater political participation, public calls for a constitutional monarchy) on the one hand, and the launching of a wave of terrorist attacks by Saudi Islamists who had previously directed their wrath at targets outside the kingdom, on the other. These were ostensibly domestic sources of change, but their external reverberations were no less worrisome. The liberalizers, often with the help of Saudi exiles, appealed to the foreign media to amplify their message, and the terrorists frequently targeted foreign installations and personnel whose departure could undermine the functioning of the Saudi economy notwithstanding the rebound in oil prices during the period under review. The most concrete response of the authorities was to revive the practice of holding municipal elections, which was suspended in the 1960s. This was a highly restricted political opening: only male citizens over twenty-one were allowed to vote, and even then only for half the roughly 1200 positions (the other half are appointed) in local councils that anyway have very little independent authority. Nor was it incompatible with the ongoing repression of liberal reformers, such as the *in camera* trial and conviction of critics demanding limitations on the power of the monarchy. Nevertheless, the very fact that the elections were held at all reflected a growing recognition on the part of at least some in the royal family that uncompromising resistance to any domestic change might well challenge the very viability of the regime.

➤ Egypt

The same calculus appears to have motivated the Egyptian government, but perhaps with even greater urgency. Unlike Saudi Arabia, Egypt has a history of political awareness and open activism that informed public life for the first half of the twentieth century before politics were anaesthetized by Nasserism. It also has a more active civil society. The stagnation that has characterized Egyptian politics for most of the past two decades, coupled with Egypt's economic malaise and growing marginality in regional and international politics, has more recently stimulated impatience for change.

President Husni Mubarak approached the end of his fourth term in September 2005 with no sign that he intended to retire and no designated successor in sight. As the prospect of six more years of bureaucratic stasis loomed larger, reformist elements became more assertive in demanding fundamental change. Protesters took to the streets under the banner of "Kifaya" (enough) and called for radical innovations such as the abolition of the state of emergency, in effect since 1981, and the introduction of truly competitive presidential elections in place of the ritualistic referendum to affirm the election of the single candidate nominated by a tame parliament controlled by the ruling National Democratic Party. Although its response also included familiar measures such as the arrest of Ayman Nour, head of the newly-formed al-Ghad ("Tomorrow") opposition party, and a round-up of "the usual [Islamist] suspects," the regime nevertheless recognized that the traditional reliance almost exclusively on repression by the security apparatus was no longer feasible. In early 2005, Mubarak therefore tried to forestall further domestic and foreign criticism by announcing a plan to hold multi-candidate presidential elections. But that concession was coupled with a provision that candidates had to be nominated by registered parties holding at least 5 percent of the seats in parliament. Since religious parties are prohibited by law, that measure effectively precluded the possibility of direct Islamist participation in electoral politics, and it prompted Muslim Brotherhood supporters to take to the streets as well. Nor was the non-Islamist opposition appeased by a measure that effectively limited participation to parties approved by the government.

As a result, a referendum called to approve the change in the election law prompted accusations of fraud and provoked even more massive public protests, brutally disrupted by attackers suspected of being thugs hired by the security forces. Furthermore, the stipulation that elections would be overseen by the judiciary, hitherto considered a central pillar of the regime, led to unexpected and unprecedented demands by judges for greater autonomy from the executive branch, without which – they claimed – they could not carry out their responsibilities. Mubarak's experience clearly demonstrates the dilemmas for authoritarian rulers persuaded by a combination of external and internal pressures to adopt a posture of defensive reform: while that may bring short-term relief, it also threatens to precipitate a broader dynamic that will eventually subvert their hold on power.

➤ Iran

For precisely that reason, the Iranian regime moved in the opposite direction. Iran had begun to experience pressures for change in the mid-1990s and actually staged a relatively open presidential election in 1997, in which the reform candidate, Muhammad Khatami, won a resounding victory. By the prevailing standards of the region, Iran then went further in terms of competitive parliamentary politics, economic and media liberalization, and women's rights. But rather than appeasing regime critics, these "concessions" by the mullahs only seemed to whet the demand for even more change, and the response was a creeping restoration of authoritarianism in all spheres of public life. That counter-offensive included brutal crackdowns on student demonstrators, progressively greater constraints on press freedoms, and restrictive rulings by a regime-controlled judiciary. Disappointment at Khatami's failure to produce greater openness during two terms in office dispirited the reformist camp and helped produce a conservative comeback in the February 2004 parliamentary elections. That, coupled with the regime's ability to mobilize nationalist sentiment under the banner of Iran's nuclear program and distribute greater economic benefits following the rise in oil prices, effectively neutralized enthusiasm for the opposition cause. When Supreme Leader Ayatollah Ali Khamenei vetoed the decision of the Guardian Council to reverse its disqualification of pseudo-reformist candidates for the June 2005 presidential election while sustaining the exclusion of real threats to the regime (along with all women candidates), critics were reduced either to a passive response – a call to boycott the election – or to reluctant support for the default option, Hashemi Rafsanjani, a former hardliner who portrayed himself as the voice of pragmatism. Neither choice proved very effective. The election was won by Mahmoud Ahmadinejad, the ultraconservative mayor of Tehran and former Revolutionary Guard. That result, tainted by insistent charges of election tampering, seemed to confirm that real transformation in Iran had, once again, become a matter for the distant future.

➤ FUTURE DIRECTIONS

With the major exception of Iran, entrenched systems all over the Middle East appear to be giving way to pressures for change, most of them generated from abroad but amplified and reinforced by internal echoes. Examples of seemingly successful outcomes of pressure for change abounded in 2004–5: demonstrations in Egypt and Lebanon, open elections in the Palestinian Authority, Iraq, and – to a much lesser extent – even in Saudi Arabia, and major policy shifts in Libya and Syria. This concatenation of events was too dramatic to be mere coincidence and many observers, seeing in it a broader phenomenon, described it as an "Arab spring," which they interpreted as a vindication of George W. Bush's policy.

Of course, not all these changes mean that the region is inexorably moving to more pacific foreign policies and/or liberal and democratic socio-political systems. If the term "Arab spring" is an allusion to the experience of Czechoslovakia in 1968, it should be borne in mind that the "Prague spring" of that year was followed by twenty more years of regression and repression. With the possible exception of Libya's foreign policy shift, all the changes discussed here are tentative, ambiguous, and perhaps easily reversed. In Lebanon, the opposition to Syria was not just an expression of democratic values. It was also, and perhaps mainly, a protest against Syria and a demand for independence that resonated differently in different ethno-religious communities, and communitarian identities may well prevent consolidation of the democratic structures and practices that already exist in Lebanon. That argument was put forward by Syrian apologists, and though it was self-serving, it was not entirely implausible. Indeed, the potential for problems of that sort was evident in the preparations for the parliamentary elections that began immediately after Syria's withdrawal and in the maneuvering that began immediately following the elections. The same challenge of communitarian identity also may overwhelm the effort to create democratic structures in Iraq. In other parts of the region, including Saudi Arabia, Syria, the Palestinian Authority, and Egypt, the most viable alternatives to existing regimes are Islamist movements, not because they necessarily reflect majority sentiment but because they are the most disciplined and most coherent in their world view. Their progress, even if achieved through democratic means, would not automatically translate into greater tolerance of cultural diversity and democratic pluralism or into more pacific and/or pro-US foreign policies.

Despite these potential complications, however, resistance to greater political and economic openness by incumbent authoritarian regimes acts more as a constraint on liberalizing and democratizing forces than on the non-democratic alternatives, and it actually contributes to the appeal of the latter. Meanwhile, it also sustains the export of dysfunction to the West. Encouragement of ongoing change has therefore been recognized as a vital interest of the West. Consequently, the West, whatever inclination it might have to disengage, will continue to be involved in the problematic exercise of encouraging liberalization in the Middle East, not just because of energy dependence but also because of the imperative to promote change in ways that will serve its own social identity and physical security.

CHAPTER ➤➤➤

2

The Ongoing Iraqi Crisis

EPHRAIM KAM

More than two years after the occupation by the US armed forces, Iraq remained marked by instability, internal struggles between and within the major ethnic groups, considerable violence, and uncertainty regarding the nature, path, and future of the country. Saddam Hussein's regime was toppled by force and will not return, but the formative characteristics of the new regime have not yet been defined, despite the efforts of the US and the various Iraqi players. Clearly the results of the crisis will not only decisively affect Iraq itself, a key state in the Arab and Muslim worlds, but will also significantly influence both other countries in the Middle East and the status and regional policy of the United States.

➤ THE ROLE OF THE UNITED STATES

The US, via the Bush administration, was the catalyst to the regime change and to the construction of a new political system in Iraq. Just as it is clear that were it not for the American military action in 2003 Saddam's regime would have continued, it is also clear that without American involvement and determination a new regime will not emerge in the country, at least in the next few years. From the outset the Bush administration hoped to present the campaign in Iraq as part of the war against terrorism waged since the September 11 attack and as a major step to halt the proliferation of weapons of mass destruction among radical regimes. However, after this picture was shown to lack sufficient basis and following the development of the grander regional vision of the administration, the campaign in Iraq has come to be regarded as part of a more comprehensive regional program.

According to this concept, and to a large extent as a result of the trauma of September 11, the US was no longer prepared to continue its traditional role as Middle East shock absorber. Instead it aimed at reforming the Middle East by encouraging democratic processes and political systems in Arab countries, introducing economic reforms to spur economic opportunities and employment options, and improving the educational systems. The Bush administration was aware that the process of change would be difficult, protracted, and gradual; would incur risks for the US and its regional allies, and was not guaranteed success. It was also clear to the administration that the major motivation for this change must come from the Arab countries themselves, and that the US and the West could play no more than a supporting role. At the same time, the administration was convinced that only a substantive change of the face of the Middle East could lead to a reduction of the hostility in the Arab world towards the US, with a consequent elimination of the terrorist breeding grounds in the Middle East and its periphery.

Iraq was regarded by the Bush administration as the focal point of the effort and a major test case for changing the face of the region. The establishment of a stable, moderate, democratic regime in Iraq would prove that this might also occur in other Arab and Muslim countries. The success of the political move in Iraq would significantly improve the international and regional status of the US, reduce the widespread criticism directed toward it (including in the US itself) for its controversial military campaign, positively affect its policy on other issues, and strengthen the position of its allies in the Arab and Muslim world. Conversely, a failure in Iraq would lead to grave results: damage to the strategic position of the US and its deterrence capability; transformation of Iraq into a locus of instability and a breeding ground for terrorist organizations; and strengthening of radical elements in the Arab and Muslim world.

The process of change has encountered many difficulties in Iraq and Arab countries and has already exacted a high price from the US, in terms of casualties, financial costs, domestic and international criticism, and increased hostility in the Arab and Muslim world. The American presence in Iraq has been regarded by many Arabs as an imperialist drive aimed at strengthening American hegemony in the Middle East and at gaining control of Iraqi oil resources, and the ongoing setbacks have further damaged its image and reliability. Nevertheless, the Bush administration has so far displayed determination to continue the efforts in Iraq, and Bush's reelection in November 2004 is hailed as approval of its Iraqi policy. Commitment to Iraq's reconstruction and rehabilitation has continued to garner considerable support in the US, and no serious elements have proposed leaving Iraq in the near future, before achieving at least an important part of the American objectives.

However, in the light of the increasing number of American casualties and the growing sense that there is no way out of the labyrinth, support for the move in Iraq has declined. Concomitantly, since the spring of 2005 there have been resonances of a public demand to outline a timetable, at least in general terms, for achievement of American objectives in Iraq and withdrawal of most of the US forces from the country. In the meantime, the US has retained a large military force in Iraq, and despite the high costs in finances and human lives, the administration was even prepared in principle to dispatch additional forces to Iraq if the need arose.

➤ RECONSTRUCTION

The Iraqi people currently face a tremendous challenge. The overthrow of the Saddam regime, the military occupation, and the efforts to build a new Iraq obligate the Iraqis to fashion a new image of their state and society and construct a regime different from what preceded it. Many difficult issues must be resolved, including: will there be a unified state or a federation? How will the power be divided among the three major ethnic groups? Is a Western democratic regime viable for Iraq? What will its orientation be on the international, regional, and Arab scenes? What will its religious identity be, and what will be the role of Islam? Will there be an Islamic theocracy? And above all, will the country retain its unity, or will it split into two or three political entities?

The resolution of all these issues will also determine the position and relative power of the ethnic groups in the future state, and they have therefore been accompanied by power struggles between the respective sectors. At the same time there also existed mutual partial cooperation, and representatives of all three groups were participating in the temporary government's institutions, both in order to influence the nature of the regime and to preserve the unity of the state.

These power struggles reflected the conflicting interests of the ethnic groups:

- For the Shiites the new regime awarded a tremendous opportunity to become the leading faction in the new Iraq, since they represent about half of the population. Internal differences of opinion and approach – between religious and secular, and between moderate elements supporting coopera- tion with the US and those opposing it – complemented other divisions regarding the nature of the future leadership, the extent of democracy, and the role of Islam. The moderate elements in the Shiite camp have thus far gained the upper hand, led by the supreme Shiite spiritual leader Ayatollah Ali Sistani, who has regarded the political process underway as the preferred path for achieving positions of power in the regime. This group therefore has preferred to cooperate with the US: even if it didn't receive all its demands, as long as the democratic process permitted the Shiites to consolidate their position and establish a regime that supported their objectives, they saw flex- ibility as in their best interest.

- Sistani has made significant progress in moderating the Shiite camp. He persuaded the young extremist Shiite leader, Muqtada el-Sadr, to reduce the violent anti-American activities of the militia under his control, the Mehdi Army, and to participate in the political process. Sadr has not halted his anti- American activities, but he has channeled them into the political field. On the other hand, since the war thousands of members of the Revolutionary Guards and Iranian agents have penetrated the Shiite concentrations in Iraq, and Iran fostered links and gained influence over various Iraqi Shiite orga- nizations. These organizations included Sadr's movement and the Shiite Badr Organization militia, originally established in the 1980s in Iran, trained by the Iranians, and directed to Iraq after Saddam's downfall.

- The Kurds have also benefited from the changes in Iraq. They threatened not to cooperate with the new regime if their demands were not met, which include a federal government, legislation to protect their interests, a secular regime, and the inclusion in the Kurdish region of the town of Kirkuk, located near large oil resources. Their demands have aroused strong resis- tance among the Sunnis and numerous Shiite leaders. In the end the Kurds succeeded in establishing the principle of a federation in the temporary constitution, thus in practice gaining autonomy in the Kurdish regions in northern Iraq – even if this principle did not advance them on the path to independence, since all the other elements, in and outside Iraq, opposed this approach. The Kurds strove for continued cooperation with the US, even at the expense of conceding some of their demands, as they wished to strengthen their influence not only in the Kurdish regions but over the entire Iraqi agenda. All in all, the Kurds have achieved impressive results. They have consolidated their autonomous government institutions; they have expanded their rule over regions that in the past were under central rule, mainly in Kirkuk, where they have consistently evicted Arab residents from the town; and for the first time they are recognized as important participants in the institutions of the central government, even beyond their percentage in the general population.

- The Sunnis, the big losers in the fall of the Saddam regime, have hoped to halt the process of reconstruction under American sponsorship because it will perpetuate their inferior status in the country. Their leadership was associated with the now defunct Saddam regime and Ba'ath party, and their current leadership has not bonded well, has produced no outstanding religious or political leaders, and has been divided among itself. Consequently a serious internal dispute has developed regarding their future in the new regime and the means to ensure their place in it, even while the extremist faction has striven to evict the US from Iraq and disrupt the establishment of the new regime by means of violence and terror. However, their ranks have also included pragmatic elements who understood that boycotting the political process was liable to harm the Sunnis and push them further away from the centers of influence. They were thus prepared to cooperate in the process, if the damage to their status was reduced and they were given a genuine role in the management of state affairs.

The foundations for the political process and for the construction of the new regime were laid mainly in the months following the war. In July 2003 the US and its coalition established the Iraq Governing Council (IGC). Most of the members of the council were Shiites, joined by some Sunni and Kurdish representatives, and while most were representatives of Iraqi exiles, included were also several prominent Iraqi personalities. The Bush administration announced from the outset that the end of the American occupation of Iraq was contingent on a new constitution and national elections for the new government, on the assumption that these would be held before the end of 2005. At the same time the American administration also determined that the IGC would be dissolved in June 2004, upon the establishment of a temporary government. However, the IGC was not sufficiently active, lacked legitimacy among the Iraqi public, and advanced very little in the formulation of the constitution because of disputes between its factions and the spread of violence in Iraq.

In November 2003 the IGC and the coalition agreed that by the end of February 2004 a temporary constitution, the Transitional Administrative Law, would be completed. By the end of May 2004 regional committees would be elected in each of the eighteen regions, which would elect representatives for the Iraqi National

Assembly, which in turn would elect a temporary government that would receive sovereignty. By the end of June 2004 sovereignty would be restored to Iraq; and by the end of 2005 national elections for a permanent government would be held.

In March 2004 the IGC ratified the Transitional Administrative Law, which determined the course for the transfer of government to the Iraqi institutions. According to this law:

- The method of government in Iraq would be democratic and federal, based on geographic and historical factors.
- By the end of January 2005 there would be an elected National Assembly of 275 delegates. The Assembly would elect a presidential council, which would consist of a Shiite president and two vice presidents, Sunni and Kurdish, and would appoint a prime minister.
- The government that would be formed after January 2005 would be responsible for drafting a permanent constitution by August 2005, which would be brought to a national vote by October 2005. Two-thirds of the voters in each three districts can veto the constitution. This decision gave the Kurds and the Sunnis, with each sector dominating three regions, the right of veto, which in turn aroused the opposition of the Shiite leaders.
- If the permanent constitution is not approved, a new version will be formulated and brought to a vote by October 2005. If the constitution is approved, elections for the permanent government would be scheduled for December 2005.
- The Kurds will set up the autonomous Kurdistan Regional Government in the areas it ruled up to the war.
- Islam is the official religion of Iraq, and will be regarded as "a source," i.e., not the sole or major authority of the constitution. No law shall be passed that contravenes the fundamentals of Islam, but neither can any law be passed that contravenes the principles of democracy and the basic rights laid down in the constitution.

Formulation of the temporary constitution and formation of the temporary government were accompanied by grave disputes between the various elements of power. The federal structure of the state, the status of Islam as a major source of the constitution, and the authority of the Kurdish and Sunni regions to impose a veto on government decisions were the major bones of contention. The opposition to these decisions contributed to the increase in violence both by the Sunnis and among Shiite Islamic elements.

In early June 2004 the temporary government was appointed under the leadership of Iyad al-Allawi and given executive authority. Many of the elements of government authority were detailed in Security Council Resolution 1546 of June 2004, which specified that the temporary government would not take long-term significant measures. Its task was to manage the government ministries and prepare the elections for the National Assembly in January 2005. As such, the temporary government would receive control over Iraq's oil income, subject to UN supervision, for at least a year. Control over Iraqi assets would also be returned to the Iraqi government. The mandate of the international force in Iraq would be reevaluated at the request of the Iraqi government, or a year after the resolution was passed. The mandate would end when the permanent government was formed, or at the request of the temporary government. Finally, the Iraqi government requested the continuation of the presence of the international force acting in Iraq, which would have the authority to take all the steps required to maintain security and stability in the country. The Iraqi government might reach an agreement with the force regarding security issues, including the management of sensitive offensive operations, and the government would have the authority to include Iraqi forces in operations of the force, after coordination and consultation with it at all levels.

In June 2004 the coalition transferred all the ministries to the Iraqi government, and on June 28 the Coalition Provisional Authority (CPA) restored sovereignty to Iraq and rule was transferred to the temporary government. The CPA was dissolved, and Paul Bremer, the American governor of Iraq and head of the CPA, left Iraq. A large American embassy, including 160 American advisors to the Iraqi government, was opened in Baghdad, and inherited some of the functions of the CPA. The American military headquarters in Baghdad became the headquarters of the multi-national force in Iraq.

Despite the return of sovereignty and power to the Iraqi government, major control in the field remained in the hands of the US, and principal elements of actual Iraqi sovereignty remained vague and unclear, particularly security and command. Moreover, since the government was not supposed to take decisions having long-term significance, it was clear that its authority was limited, even if it officially assumed control of the oil revenues. As such, the actual important element in Iraq remained the US, even though it had agreed to formal and practical limitations of its authority. At the same time, US influence in Iraq was no longer based solely on dictates, but also on the capability of persuasion, supported by economic and military aid.

The next important stage in the construction of government institutions were the elections in January 2005 for the 275 seats in the National Assembly, and for the regional assemblies in the eighteen regions of Iraq and for the regional assembly in Kurdistan. The overthrow of Saddam's regime and the preparations for the elections created in Iraq unprecedented freedom of political organization. In the elections themselves, voter turnout, 58 percent, was higher than expected. This was regarded by the Americans as a success, since they considered the elections to be an important step in the consolidation of the regime in Iraq and in the expansion of democracy in the Middle East.

The elections were also deemed a success by the Shiites and the Kurds, who felt that they gave legitimacy to the change in relative forces in Iraq and to the end of the Sunni domination over the Shiite majority. However, the elections themselves reflected a serious domestic problem. The overall rate of voting was high because 70 percent of the Shiites and 90 percent of the Kurds participated, but the percentage of voters among the Sunnis was minimal (2 percent), and only seventeen Sunni delegates were elected to the National Assembly. The low level of participation by the Sunnis reflected their frustration with the new regime and their expulsion from the centers of power, as the new regime had already previously begun pushing the Sunnis outside the political system, This in turn motivated them to take extreme steps in order to undermine the political process.

The need to prevent the alienation of the Sunnis was clear to everyone. Iraq cannot have a stable regime if a fifth of the population opposes it. The leaders of the largest Shiite group in the parliament declared their intention of appointing Sunnis to positions in the government and to the committee that would formulate the permanent constitution. They promised that the country would not be run by religious figures and dictated by Islamic law. Important Sunni organizations have expressed interest in such conciliatory steps, but they presented rigid conditions for their participation in the process, including a timetable for the withdrawal of American forces.

In the spring of 2005 signals were also received from radical Sunni leaders that they would be prepared to enter the political process if they were given a viable role in running the country. The Americans pressured the Shiite leadership to make concessions to the Sunnis, in the hope that it would be possible to isolate the extremists by the integration of more moderate Sunni elements in the political process. As a result, Shiite leaders and the provisional Iraqi government displayed readiness to appoint Sunni personalities to senior posts in the administration and in the constitutional committee. The head of the provisional government even proposed that members of the Ba'ath party not tied to crimes of bloodshed could participate in the construction of the new regime, and that amnesty would be granted to rebels who were not involved in violence and who would be prepared to lay down their weapons.

Despite the indications of an incipient trend within the Sunni leadership to participate in the political process, numerous difficulties remained. As far as the Sunnis were concerned, the decisive issue was whether they would be capable of overcoming internal disputes and acting as a cohesive political bloc. Neither was it clear to what extent the Sunnis would abandon their basic demands and become accustomed to their inferior status in the new regime, after being the dominant force in Iraq for generations. Not less important but as yet unclear was the degree of influence of the Sunni personalities willing to join the political process, including groups of insurgents who were divided and lacked central leadership. On the other hand, although Shiite leaders understood the need to include the Sunnis in the regime, after many years of oppression, they were eager to establish their new leadership status and cleanse the administrative organization of Sunni elements and Ba'ath party veterans. Consequently many of them were unwilling to share power with the Sunnis, and considerable mutual suspicion remained.

In late April 2005 the political process advanced with the National Assembly election of a new government, headed by Ibrahim Jaafri. The government comprised thirty-seven ministers: a Shiite prime minister, four deputies (three of them representatives of the large ethnic groups), and thirty-two other members, including seventeen Shiite ministers (among them three ministers from the political movement of Muqtada al-Sadr, which earned twenty-three delegates to the National Assembly), eight Kurds, five Sunnis, a Turkoman, and a Christian. At the same time the Kurdish leader Jalal Talabani was elected president of Iraq – reflecting the increasing weight of the Kurds in the new political system – with two deputies, a Shiite and a Sunni.

The election of the government was couched in tension, manifested by the absence of about a third of the Assembly delegates during the vote and in delays to the appointment of ministers. Sunni alienation from the political process was expressed not only in the refusal of prominent Sunni personalities to participate in the government, but also in that the influential committee appointed to formulate the permanent constitution by August 2005 included only two Sunnis among its fifty-five members. Only after

American pressure on the Iraqi government was a compromise reached, in which the number of Sunni committee members was increased to fifteen.

At the ruling level, the Jaafri government is mainly built on the two largest parties in the National Assembly: the Shiite United Iraqi Alliance, with 140 out of 275 seats, and the Kurdish Alliance, with 75 seats. The division of power between the Shiites and the Kurds is also reflected in their agreement to continue to maintain the major independent militias of both ethnic groups. Special importance was given to the agreement to permit the continued existence of the Peshmerga Kurdish militia. These developments reflected the fact that the change in regime in Iraq also involves a change to the elites dominating the country. The new regime is based on a Shiite–Kurdish coalition, with the Sunnis examining their place in the system, cognizant that any political arrangement will deny them the extensive power that they had previously enjoyed.

➤ **ECONOMIC REHABILITATION**

In addition to construction of the new regime, the US has devoted intensive efforts to rehabilitate the Iraqi economy, seriously damaged by Iraq's three wars over the last twenty-five years: the Iraq–Iran War of the 1980s; the 1991 Gulf War (and the sanctions subsequently imposed on Iraq); and the 2003 Iraq War and the ensuing occupation. The average income capita in Iraq dropped from $3600 in 1980 to $770–1020 in 2001, and to $450–600 at the end of 2003. The Iraqi economy suffered from additional problems in the period 2004–2005: a high rate of inflation, serious unemployment, low purchasing power, and disruption of economic activities because of violence and problems of internal security.

It was clear to the American administration that in the construction of the new Iraq, rehabilitation of the economic system was essential for stabilization of the future regime and acquisition of trust and cooperation on the part of the population. The major expenditures on rehabilitation were directed at establishing security and order, at training and equipping the Iraqi security forces, and at infrastructure programs – electricity, water, oil production and export, and road construction. As part of the rehabilitation plan local councils and trading centers were established, clinics, hospitals, and thousands of schools were reopened, courses in democracy were offered, and encouragement was given to the private sector.

In October 2003 the World Bank and the UN estimated that $36 billion was needed over four years in order to rehabilitate the fourteen major sectors in Iraq. To this figure was added $19.4 billion that would be required, in the coalition's estimate, for the needs of security and the rehabilitation of the oil and other sectors. The overall cost of rehabilitation to the end of 2007 was therefore estimated at $55 billion. The financial aid was to be received from the US, from international donors, and from Iraqi sources, mainly in oil revenues. In April 2003 the US allocated $2.5 billion for the rehabilitation of Iraq, and in November 2003 allocated more than $18 billion – mainly for maintaining security and for the rehabilitation of the electricity, water, and drainage systems.

However, the violence and other difficulties did not allow for spending all the allotted funds. In the first year after the war only $2.2 billion were actually spent, out of the more than $18 billion allocated by Congress. The rate of spending has significantly increased since November 2004, as a result of the initiative taken by the American administration to speed up the rate of rehabilitation. Still, the surplus made it unnecessary to request from Congress additional rehabilitation funds for 2005. However, the administration did request a further allocation of $5.7 billion for the training and equipping of the Iraqi security forces. To this were added promises that various countries (headed by Japan and Britain) gave in October 2003 to allocate $3.6 billion as a grant for the rehabilitation of Iraq, and another $13.3 billion in the form of loans.

The American administration has argued that despite the violence, ongoing progress has been achieved in economic rehabilitation: life has returned to normal in most parts of Iraq; the economy is recovering; the electrical system has once again begun functioning, and at a level higher than before the war, despite the terrorist attacks; the level of sanitation, health, and education has improved; and a new currency was introduced and has remained stable since the beginning of 2004. At the same time, rehabilitation efforts have encountered great difficulties, and violence, terrorism, and security problems have significantly impeded the progress. Some projects have been characterized by lack of efficiency, while rehabilitation was also slowed by additional causes: disputes between American administrative bodies regarding control of allocations; bureaucratic difficulties and procrastination on planning; unrealistic application of American concepts to Iraqi conditions, such as emphasis on privatization; and reservations on the part of Iraqi elements regarding the execution of plans, because of the fear of American procedures. As a result, only a few of the rehabilitation programs were implemented, and objectives were not achieved in critical sectors, such as electricity and oil.

A critical link in the rehabilitation effort was the oil sector. Before the war the American administration expected that Iraqi oil revenues would help to finance a large part of the rehabilitation costs, if not all of them. The Security Council resolution of May 2003 that ended the sanctions imposed on Iraq following the Gulf War authorized the coalition to use the oil revenues for purposes of long-term rehabilitation. The resolution also transferred the responsibility for oil profits from the UN to the US, by the establishment of the Iraqi Development Fund.

Yet these hopes were only partly realized. The Development Fund monies aided in the financing of various rehabilitation activities, but thereafter the difficulties facing the rehabilitation of the oil industry increased. Although the oil industry was not seriously damaged in the war, it became a major target for terrorist attacks and robberies after the war. Oil exports were renewed in June 2003, and gradually increased to 1.3 million barrels a day in April 2004, compared to 2.2 million barrels before the war. By October 2004, oil exports still did not exceed 1.4 million barrels a day, although overall oil production in September 2004 reached a peak of 2.67 million barrels a day – greater than the pre-war level of 2.5 million barrels. In February 2005 production dropped to 2.1 million barrels a day, although the target for the end of 2004 was 2.8–3.0 million barrels a day, and the long-term target was 4–6 million. Apart from the depressed state of the oil industry, which desperately required development and modernization, the major problem regarding the increase of production and export was the terrorist attacks on major oil pipelines and installations that disrupted the transport and export of the oil.

➤ THE SECURITY SITUATION

The gradual and partial successes of regime construction and rehabilitation of the economy have been overshadowed by the internal security situation in Iraq. The violence and the terrorist attacks began a short time after the occupation of Iraq, but escalated mainly from the end of 2003 and the beginning of 2004. The terrorist attacks have been perpetrated primarily by two groups. The major one is apparently composed of Sunni elements, some of them supporters of Saddam Hussein and members of the Ba'ath party. Essentially they have been unwilling to accept the reality of the new regime, in which the leading Shiites in cooperation with the Kurds have already ousted the Sunnis, for generations the mainstay of the regime. The terrorist attacks were thus intended to harm the American forces and prompt their departure from Iraq; remove foreign elements from Iraq; and stir up the local population against them, because without the US force behind the new regime, it will collapse. At the same time the terrorist attacks were directed no less against Iraqi elements cooperating with the US or playing a role in the new regime; the emerging institutions of the regime; and the economic infrastructure, in particular the oil installations.

The second group comprises various Islamic elements, including Shiites motivated by hatred for the US and opposition to the American occupation, and driven to exploit the vacuum created in Iraq and attempt to establish an Islamic regime. One of the prominent elements in this group was Muqtada al-Sadr, who threatened in the spring–summer of 2004 to arouse a general uprising against the Americans in a group of towns with a Shiite majority. Sadr's Mehdi Army ultimately failed to lead the general uprising, despite his influence in several of the towns in the south, and in the face of the obstruction by the Americans and the casualties sustained by his forces, he was forced to come to several compromise agreements in the second half of 2004. Hence – and following attempts at persuasion by the Iraqi government – his subsequent declaration of interest in joining the political process and his success in the elections and the government ministries.

Iraqi insurgents were joined by many outside Islamic fighters who entered Iraq, mainly from Syria and Iran. These included members of al-Qaeda and similar organizations who, in a drive to expel American forces and foreign elements from Iraq, regarded them as a convenient target for terrorist attacks. Moreover, there has been a prominent increase in Islamic activities in Iraq after the war, both Shiite and Sunni. Basra, for example, the second largest city and overwhelmingly Shiite, has become a focus of Shiite extremism; and Fallujah, in the heart of the Sunni triangle, has become a bastion of Sunni Islamic elements. August 2003 saw the beginning of a wave of unceasing terror against a long series of targets: US and coalition forces; Iraqi security forces and the recruiting and training facilities prepared for them; Iraqi personalities and government institutions; and embassies, hotels, and employees of foreign companies. Up to the end of July 2005, about 1650 American troops were killed and more than 12,000 injured in these terrorist attacks, not including the 140 American soldiers killed in the war itself; the more than 300 killed in accidents; and the 200 coalition troops killed. However, the majority of the attacks (and the suicide bombings in particular) were directed at Iraqis: more than 20,000 Iraqis were killed, including a large number of police, but with the majority innocent civilians.

The picture of American casualties is not constant over time, as charted by figure 2.1. Since May 2003, the American forces suffered on average more than sixty fatalities a month. In most months the number of fatalities fluctuated between thirty and eighty, but in peak months (April and November 2004) fatalities exceeded more than 130 fatalities per month. And while as yet there has been no substantive decline, there has been an increase in attacks on Iraqi security forces and against the Iraqi population. Indeed, American officers estimated that over time the operational capabilities of the insurgents, who number at least 20,000 fighters, have improved. They have begun operating in larger units, their operations are better coordinated; they have established strongholds in several regions; their capability of persisting in battles has improved; and they have developed a better capability of waging organized resistance. As a result of the activities of the insurgents, considerable parts of Iraq lay, at least temporarily, outside the control of the American and Iraqi forces.

Confronting the insurgent organizations in 2005 were 140,000 American troops, as well as 50,000 US soldiers in Kuwait in supporting and relief roles. The other coalition forces totaled about 25,000 troops from twenty-five countries, most of them from Britain and Poland. Overall, the international distribution of the security forces in Iraq is problematic. The American administration made great efforts to include peacekeeping forces from additional countries, in order to overcome security problems in Iraq and award international legitimization to its activities there. However, this participation has been limited because of the opposition of many countries to the war and their reservations regarding the US activities in Iraq. NATO agreed to train Iraqi soldiers and supply logistical support to the international forces in Iraq, but was not prepared to send forces there, despite the request of the US. Several countries that sent forces to Iraq removed or reduced them following terrorist attacks.

The US administration strove to build Iraqi security forces that ultimately would ensure the stability of the country, but they have proven even more problematic. Though technically under the authority of the Iraqi prime minister, the forces have been shaped by American embassy in Baghdad and the American military command. At the beginning of 2005 they comprised about 55,000 trained Iraqi police, with the goal of 135,000 police by the middle of 2005. The former massive Iraqi army was dissolved, and the US intended to create a small army of 27,000 soldiers by the middle of 2005. Members of the former paramilitary National Guard were incorporated into the army, and together with the military forces, numbered 57,000. To these should be added various small forces, including the response force and the Border Guards. The Iraqi security forces thus totaled 160,000 in the middle of 2005 – about half the number originally intended.

The Americans have experienced substantial difficulty in increasing the Iraqi forces as planned and in improving their quality. The Iraqis' fear of terrorist attacks, numerous desertions, penetration of the ranks of the recruits by the insurgents, involvement in crime, and a shortage of equipment and training facilities have been the principal challenges. While the Iraqi forces by the summer of 2005 were organized into 107 battalions, according to the US military only three of them so far were capable of carrying out independent roles, twenty were capable of carrying out tasks in cooperation with the American forces, and the remaining battalions were still in training. By mid-2005, therefore, the Iraqi forces were assigned tasks of reconnaissance, manning roadblocks, and guarding installations, and only a few units were sent on

US Soldiers Wounded in Iraq

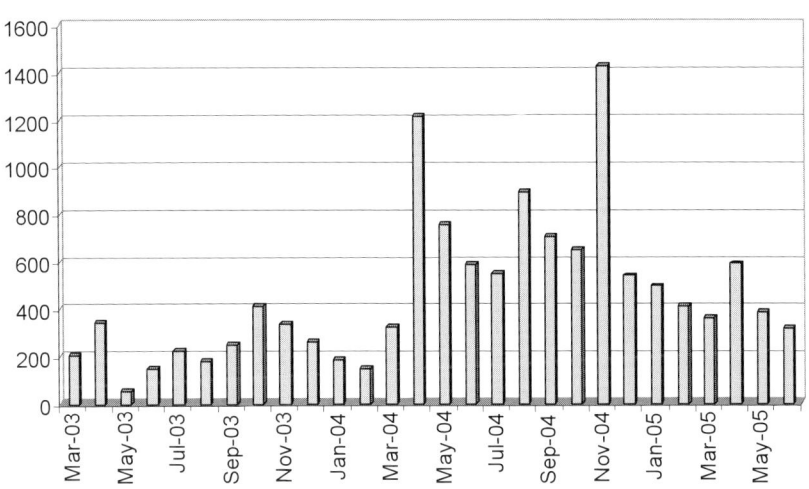

US Soldiers Killed in Iraq

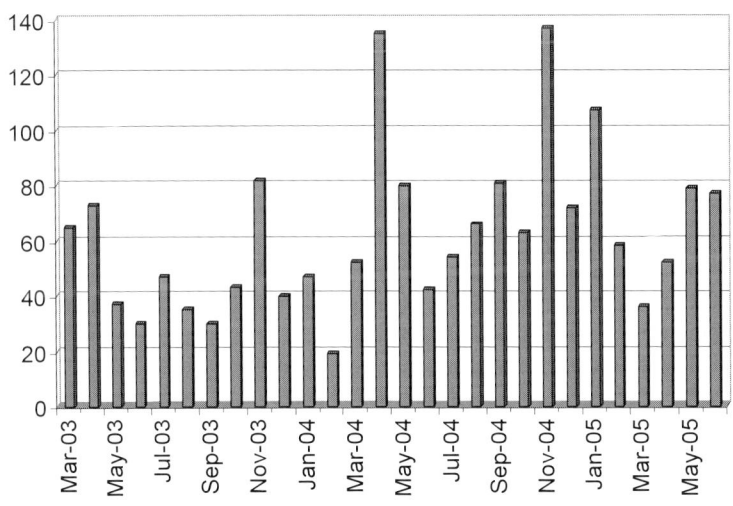

Figure 2.1   US Casualties in Iraq, May 2003–June 2005

Data source: <www.globalsecurity.org/military/ops/iraq_casualties.htm>.

offensive missions. Their performance was poor, and they generally avoided fighting against the insurgent forces. Over time their capability has improved to a certain extent, and some units have fought against insurgents when operating alongside the American forces. Nonetheless, in the middle of 2005, American officers estimated that the Iraqi security forces were incapable of maintaining stability and security in Iraq, and that at least two years would be required until enough Iraqi forces were formed and trained to permit large numbers of American forces to withdraw from Iraq.

The Iraqi police force has become a major target for terrorist attacks, mainly in the Sunni regions, and since they operate within the population, the police were particularly exposed and vulnerable to attack. By the end of 2004 more than a thousand policemen were killed. The attacks on the police were intended to undermine the structure of the regime and its stability, to highlight the absence of security in the streets, and to strike at the government's efforts to curb the activities of the insurgents. The attacks have not only caused serious casualties but have also damaged morale among the police, now in need of support from the coalition forces for the force's rehabilitation. The Iraqi security forces in turn have encountered other grave difficulties, including the recruitment of manpower, mainly experienced officers, the acquisition of accurate intelligence, and management of a complex political environment.

The attacks by the insurgents have also significantly affected the security situation in the area. They have forced the coalition forces to change their deployment, have led to the withdrawal of foreign forces, international organizations, and economic corporations from Iraq, and have weakened the readiness to participate in the coalition. They have also forced the US and Iraqi governments to devote tremendous efforts to defense against terrorist attacks instead of to the rehabilitation of the country. On the domestic front, the insurgents have terrified the population in certain regions, and prevented it from cooperating with the government and the American forces. For their part, the coalition forces have encountered difficulty in coping with the terrorist attacks. Lack of high quality intelligence – limited because of unfamiliarity with the Iraqi system, language problems, shortage of experienced intelligence personnel for rapid replacement, difficulties in penetrating the insurgents' organizations and acquiring human intelligence, and lack of sufficient cooperation on the part of the population – has hindered the struggle against terrorist attacks.

The US forces from time to time have carried out larger operations against the insurgents, mainly in order to destroy their strongholds in urban regions, but also in order to block their penetration routes into Iraq, in particular via the Syrian border. Despite various difficulties, these operations have generally ended in at least partial military success. The Americans have tried to assign to the Iraqi forces the task of maintaining security in the region of the operation after its completion, but with inadequate results. Nor have the Americans been very enthusiastic about operations involving extensive destruction in urban areas – as in Fallujah in November 2004 – because such operations seriously harm the local population and intensify the hatred for the US. In all, the difficulties encountered in conducting the operations, guarding the "cleansed" area, and sealing its boundaries have led American officers to argue that the scale of American forces operating in Iraq is insufficient to carry out the tasks assigned to them.

The unstable security situation has had a direct effect on the rate and extent of Iraqi's rehabilitation. The terrorist attacks have substantially increased the cost of the rehabilitation, and the sabotage against the oil pipeline has caused a loss of billions of dollars in oil revenues. Foreign employees involved in economic projects have left Iraq because they feared for their lives, there is a severe shortage of foreign experts, and companies are therefore avoiding investing in Iraq. Projects have become far more costly because of protection and security costs, employee insurance, power cuts caused by terrorist attacks, and poor management due partly to a shortage of skilled manpower.

➤ BUILDING THE REGIME AND THE STATE: FUTURE CHALLENGES

The process of building a new Iraq so far suggests a mixed balance of opportunities and risks. Regarding the opportunities, four important elements should be emphasized:

First, the process of building the regime is continuing more or less in accordance with the model designed by the American administration, despite the delays, difficulties, and disputes between sources of power in Iraq. The temporary constitution and temporary government institutions, including a National Assembly, a Presidential Council, and a government, were all critical milestones. The political process was expected to advance to the formulation and ratification of a permanent constitution and the election of permanent institutions. No less important, the Iraqi establishment of security forces that are to restore security and stability to the country has begun.

Second, obstacles and impediments notwithstanding, the economic infrastructure has begun a gradual rehabilitation. Oil production has approached the pre-war level, although still far from the target figures set up by the US, and oil exports too have lagged far behind the pre-war level and their target capacities because of sabotage and terrorist attacks. The electrical system has almost been restored to its pre-war level, despite a frequent energy shortage and disruptions to the working of the system. Considerable progress has also been made in rehabilitation of the educational, health, and sanitary systems and the operations of municipal government.

Third, most Iraqi groups are interested in American and international forces remaining in Iraq and continuing the regime and state reconstruction. No particular affinity for the United States drives this sentiment, rather the understanding that should the forces withdraw prematurely, the entire system is liable to collapse and succumb to anarchy, if not civil war and a dissolution of the state. Most Shiite and Kurdish leaders are also interested in a continued American presence since the regime currently under construction suits their interests.

Fourth, the Bush administration, with the breathing space it acquired through reelection, has displayed its determination to continue its efforts in Iraq, costs, losses, and criticism notwithstanding. There is insufficient domestic pressure to prompt an immediate withdrawal from Iran, and even international voices that condemned the American campaign and objected to the American administration of Iraq have evinced the awareness that a premature withdrawal of American forces is liable to leave a dangerous vacuum. As a result, international forces have displayed willingness to contribute to reconstruction efforts, even if to a limited degree.

Challenging these positive signals are several significant dangers to the future political process in Iraq, chief among them terrorism, violence, and the lack of security. Indeed, these dangers eclipse many of the achievements gained thus far. The lack of security impedes effective government, obstructs economic reconstructive efforts, challenges the authority of the United States, and threatens to prompt a premature withdrawal of foreign forces. Thus far, Iraq and the United States have not brokered a stable security situation, and without the requisite internal security, any gains scored are in serious jeopardy. Moreover, the grave difficulty in removing the American forces from the country before the minimal objectives of regime construction are achieved has begun to create the impression that the US is mired in Iraq with no way out. Such a situation cannot continue indefinitely. If no real change occurs in this situation within the next year or two, the American administration is liable to face the need to give up some of its objectives – such as building a democratic regime in Iraq – and come to terms with the characteristics of a regime that are, from the United States' vantage, less than satisfactory. Growing domestic pressure might force the administration to shorten the timetable of the forces' deployment in Iraq.

Second, while regime institutions have begun to emerge, some are at best organizational façades that barely veil blatant ethnic rivalries. The American presence has encouraged compromise and joint efforts toward progress, but the cooperative effort and its ability to spawn and sustain a moderate, democratic regime once American forces withdraw is highly questionable. In the meantime, the conflicting interests of the respective ethnic groups, the lack of any democratic tradition, the Islamic and clannish elements embedded throughout Iraqi society, and the fact that most of the region supports neither the democratic efforts nor the rise of the Shiites and the Kurds – and the consequent implications for their own regimes – are all factors that cast the construction of a stable democratic regime in Iraq into question.

Third, the drastic and rapid change to the internal balance of forces in Iraq has created additional risks. Shiite domination is liable to lead to enhancement of the Islamic characteristics of the regime and to an increase of Iranian influence in Iraq. The American administration has already voiced concern that religious Shiite groups will use their strength in the government institutions to build up Iraq as an Islamic state.

Fourth, the economic situation in Iraq remains severe because of the recent war damage, particularly in context of the previous difficult twenty-five years. It is hard to envision serious progress in the building of the country when the standard of living is only 12–15 percent of the 1980 level, when the unemployment rate is close to 50 percent, and when there is such high inflation and such low purchasing power.

Fifth is the question of Iraq's future as a united country. The federal structure of the state is a fait accompli, although differences of opinion exist regarding its form and the division of forces and authority. The Kurds have already established their status as an autonomous unit with semi-independent characteristics. Some Shiite leaders also think that in the light of the difficulties involved in Sunni participation in the political process and the unceasing violence in the country, it would be better to establish an autonomous Shiite region in central and southern Iraq. In such a case Iraq would become a federation having three strong governments as well as a central government. The continuation of the hostility and violence between and within the ethnic groups that have separate armed militias is liable to lead to greater fragmentation of the Iraqi system.

In the final analysis several factors will determine the success or failure of the political process in Iraq. First and foremost is the capability of halting the wave of violence and terror and of stabilizing internal security. In this context, the American capability by itself is apparently not enough, and the key will mainly lie with the Iraqis. Success will hinge on convincing Iraqi elements involved in terrorist activities to abandon this course and participate in the construction of the new regime. It will also depend on the capability of the US and Iraqi governments to build up effective Iraqi security forces that will gradually assume most of the responsibility for internal security. At the same time, the construction of Iraqi security forces also demands caution, in order to prevent their becoming an internal political force liable to try to gain power after the US withdraws from Iraq.

Prominent Sunni leaders must be persuaded to encourage their followers to participate in the political process and abandon the use of violence. This will obligate the Sunnis and Kurds and the Americans and international bodies to enact a series of confidence-building measures that will guarantee the Sunnis a reasonable role in the future regime. The American administration already understands that military operations by themselves will not halt terrorism and that political moves are also required with the aim of isolating the terrorist elements.

As long as the violence continues, there must be a strong capability of waging a sophisticated campaign against the insurgent organizations. This campaign obligates careful use of force in order to minimize harm to the population, and the use of political and economic moves that will help to stabilize the regime, improve the economic situation, and isolate the insurgents. An essential condition for waging the military campaign will be the capability of acquiring accurate, high quality intelligence about the insurgents.

Additional international bodies that will aid in stabilizing the situation must be enlisted. This need is already clear to the American administration, and may obligate concessions regarding its influence and control in Iraq, in order to persuade others to lend their support. At the same time, much depends on the determination of the US to continue its efforts and leave most of its forces in Iraq. The question is whether the resolve of the Bush administration to continue its efforts will slacken if the price spirals – for example, as in the event of a massive terrorist attack – or if the impression is strengthened that the efforts in Iraq aren't leading anywhere and harbor few chances of success.

In addition, the major regional implications of the Iraqi crisis, first apparent immediately after the war, continue. Iraq has disappeared as a military force, at least for many years, and its economic and political strength in the region is also limited. The Iraqi threat has thus been removed from various countries, including Iran and Israel, for a considerable period of time. The American military move in Iraq has also created tremendous pressure on the neighboring radical regimes of Iran, Syria, and Libya. Iran has been forced to maneuver and make temporary concessions regarding the nuclear issue in the face of increasing international pressure, in part because of the fear of economic sanctions, and perhaps because of the fear of a limited military campaign. Libya has changed its strategic position and abandoned its involvement in terrorism and attempts to develop weapons of mass destruction. Syria damaged its relations with the US, in part because it permitted terrorism operatives to travel to Iraq through its territory, and in the face of increasing international pressure, it has been forced to make difficult concessions on its Lebanese interests. Finally, the US has underscored its status and capability of action as the sole superpower. However, its sinking in the Iraqi swamp has highlighted the limitations of its force, raised questions regarding its reliability and power of deterrence, and increased hostility in the Muslim and Arab world.

At the same time, two potential developments following the Iraqi crisis have not materialized. Although the Kurdish sector in the Iraqi system has consolidated its status as a weighty autonomous factor, this development has not yet created a significant wave of fermentation and instability in the Kurdish sectors of Iraq's neighbors – Iran, Turkey, and Syria. This is not necessarily the last word, and such a development is still possible, particularly if the Kurdish region in Iraq takes further steps towards independence. In addition, Iraq has not yet become a producer of terrorism for other countries, and thus far terrorist activities are imported rather than exported. Here too the final chapter has yet to be written, since the focus of the terrorist organizations remains the American forces and the bodies aiding them in Iraq. It is still possible that at a later stage terrorist groups will direct their efforts at objectives outside Iraq, as occurred in Afghanistan. This, however, would most likely develop after most of the American forces have been withdrawn from Iraq.

CHAPTER ➤➤➤

3

Israel and the Palestinians: A New Reality under the Shadow of Ongoing Conflict

SHLOMO BROM

The period since early 2003 has been characterized by several important developments that may exert major influence on the future of the Israeli–Palestinian conflict, as well as on chances of resolving the conflict, or at least diminishing its intensity.

The first significant development, which began prior to 2003, was the gradual fading of the armed intifada, at least in the West Bank. The second development was the slow progress on construction of the separation barrier (the fence) between Israel and the West Bank. The third development was Israel's initiation of the unilateral disengagement plan and the steps toward its implementation. The fourth development was the death of Palestinian leader Yasir Arafat and the emergence of a new and different Palestinian leadership. A fifth noteworthy development, which took place outside the local arena but which may also have a significant impact on the region, was the reelection of American president George W. Bush.

These developments have created an opportunity for positive change in the dynamics of Israeli–Palestinian relations. However, they exist alongside longstanding elements integral to the Israeli–Palestinian conflict, including a mutual lack of trust, significant gaps on major issues, and new trends within Israeli and Palestinian societies that not only make actualizing this opportunity difficult but threaten to fuel and intensify the crisis.

➤ THE DECLINE OF THE ARMED INTIFADA

Operation Defensive Shield, conducted by Israeli forces in the West Bank between March 29 and May 10, 2002, was a turning point in Israel's war on Palestinian terrorism. During the operation, the IDF reoccupied the areas of the West Bank that had been under Palestinian control and badly damaged the capability of armed Palestinian groups to resist the entry of Israeli forces into all areas, including densely populated sites such as Palestinian refugee camps and the Nablus Casbah (Old City).

Israel did not leave a permanent military presence in heavily populated Palestinian urban areas, but from this juncture onward Israeli forces enjoyed freedom of movement and freedom of action, which facilitated the acquisition of better intelligence and the more effective elimination of terrorism cells. Thus, even if the operation inflicted only minor short-term damage on the ability of Palestinians to carry out painful attacks against Israel – primarily suicide attacks, which continued to be executed immediately following the operation – it also resulted in a gradual improvement in Israel's ability to thwart attacks and to strike at the forces in the West Bank initiating them. These achievements were reflected clearly by decreasing Israeli death tolls since 2002, as represented in figure 3.1.

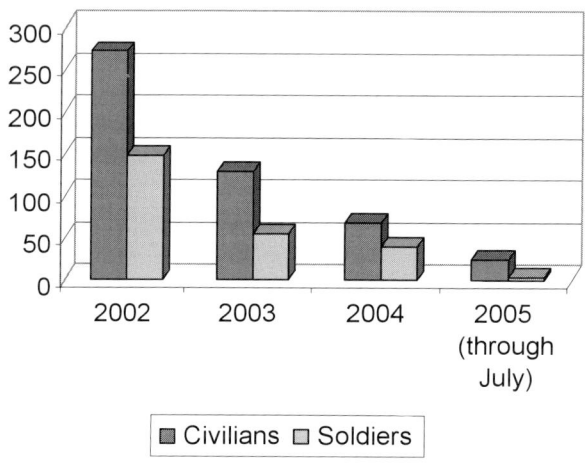

Figure 3.1   Israeli Fatalities in the Israeli–Palestinian Conflict, 2002–2005

Data source: <http://www.btselem.org/Hebrew/Statistics/Casualties.asp>.

Most home-front casualties of the intifada years were caused by suicide attacks. Israel's increased ability to interrupt suicide attacks is charted in figure 3.2, which compares the number of suicide attacks executed with the number of planned attacks that were derailed or not completed for other reasons.

The situation was different in the Gaza Strip. From the outset, Israel was less vulnerable to terrorist attacks from the Gaza Strip because of the highly effective fence constructed around the area at the beginning of the Oslo process. Thus, the number of successful attacks carried out in Israel that originated in the Gaza Strip was already very small. This lower level of vulnerability enabled Israel to adopt a different operative strategy against Palestinian violence in the Gaza Strip. Israel refrained from re-conquering the territory under Palestinian control, and instead relied on defensive military deployment and pinpoint strikes at armed groups, usually from a distance. In the event of more intense outbursts of violence, Israeli ground forces carried out incursions into areas on the periphery of Palestinian controlled territory. From mid-2003 onward, Israel also employed pinpoint strikes against the leadership of Hamas, the main Islamic armed movement, on the assumption that any distinction between the group's civilian and military leadership was artificial and that they were actually one and the same. The Israeli operations did not cause direct serious damage to the operational capabilities of armed groups in Gaza, but rather demonstrated to them the price of their activities. At times, these IDF operations forced those initiating attacks to focus on going underground and trying to survive, resulting in their having less time and fewer resources to launch operations.

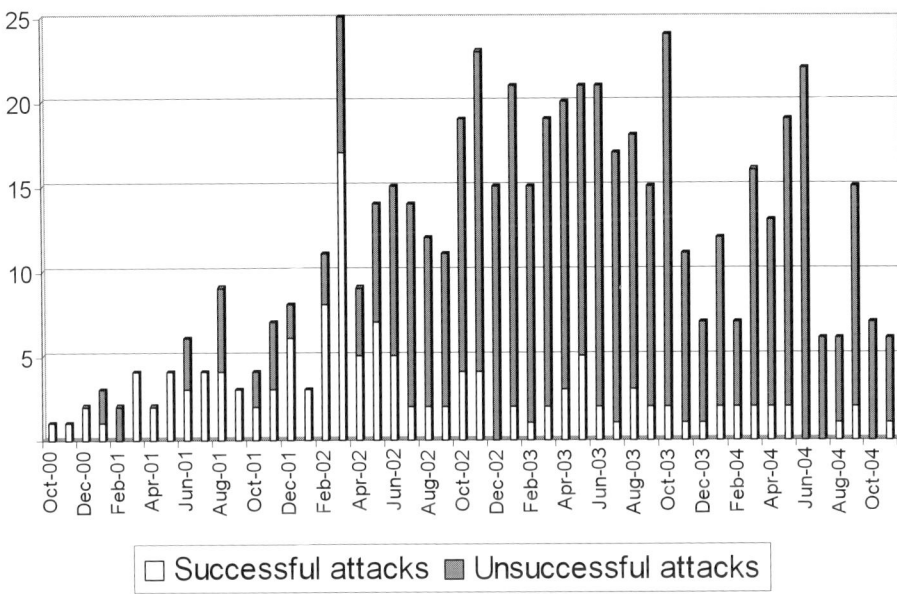

Figure 3.2 Suicide Terror Attacks, Successful vs. Unsuccessful (thwarted), 2000–2004
Unsuccessful (interrupted / thwarted attacks): preventative operations of the General Security Services and IDF, "work accidents," mishaps during execution, and so on.

Total successful attacks: 138   Total unsuccessful attacks: 431

Data Source: <http://www1.idf.il/SIP_STORAGE/DOVER/files/6/31646.doc>.

In addition to the diminishing achievements that could be attributed to terrorist activities, the harsh blow to the armed groups – reflected in the large number of casualties and arrests – strengthened those Palestinians who objected to the militarization of the second intifada. This included those who had expressed their opposition to the use of violence from the outset, the most prominent of whom was Mahmoud Abbas (Abu Mazen), as well as others who joined the opposition upon realizing that violence was not effective and was actually harming the Palestinians more than it was harming Israel. Even when Arafat was still alive, the growing strength of this platform resulted in public expressions of opposition to the use of violence and the Palestinian leader's policies, and officials within the Palestinian Authority (PA) were no longer as frightened to express their positions openly.

➤ SEPARATION AND DISENGAGEMENT

In April 2002, the Israeli government approved the construction of a barrier ("the separation fence") around the West Bank. Implementation was delayed at various points by political decisions on changes to the fence's route and the legal problems involved with demarcating a barrier that did not sufficiently consider the needs of the Palestinian population. On July 9, 2004, the International Court of Justice in The Hague generated an advisory opinion that construction of the barrier was illegal. Israel, however, refused to cooperate with the court, rejected its decision, and held that the court was mistaken in its ruling.

While construction of the barrier around the northern portion of the West Bank and parts of Jerusalem was completed by mid-2005, the barrier was not expected to be completed before the end of 2006. Already, however, the fence has proven itself as an effective barrier to the penetration of attackers into Israel. At the beginning of the second intifada, cities in northern Samaria (Nablus and Jenin) were the major source of suicide attacks in Israel. Completion of the northern section of the barrier imposed on attackers a much greater distance to Israel, which provided the Israeli security forces many more opportunities to intercept would-be terrorists while still en route.

The construction of the barrier also embodied political significance, especially as it established in Israeli public consciousness the understanding that there is no hope that the Israeli settlements located on the Palestinian side of the barrier will ultimately remain in Israeli hands. Moreover, despite efforts by Israeli officials to stress that the considerations underlying the barrier's route were entirely security-related, that it was not a border, and that it carried no political significance whatsoever, the fence began assuming the status of a border between Israel and the Palestinian territories. As a result, from the outset settlers battled over the route of the barrier and the inclusion of different settlements to its west; the two most prominent examples were Ariel and Ma'ale Adumim (figure 3.3). The government decided to include both of these cities within the route of the barrier despite the strong opposition of the US government, which regarded the move as creating facts on the ground that would interfere with future solutions to the conflict. While American opposition did not cause Israel to change its decision, Israel did suspend the process of connecting the portion of the barrier adjacent to these cities with the barrier's main route. The construction freeze in this area was based on an assumption that it would be possible to connect the respective segments in the future, when suitable political conditions arise.

Opposition to the route of the barrier for its interference with the normal lives of the Palestinian population reached the Israeli Supreme Court. In July 2004, in an important landmark decision, the Court ruled that the government was required to strike the right balance between security considerations and legitimate Palestinian demands for a normal fabric of life, as required of Israel by international law. This Supreme Court decision led to the re-demarcation of several sections of the barrier, which usually served to bring it closer to the Green Line. According to mid-2005 plans for the route of the barrier, only 9 percent of the West Bank was to be located on the Israeli side of the fence. In the Palestinian camp, there was common solid opposition to the construction of the barrier, deemed a unilateral step that aimed at Israel's de facto annexation of Palestinian territory. The Palestinians argued that the construc-

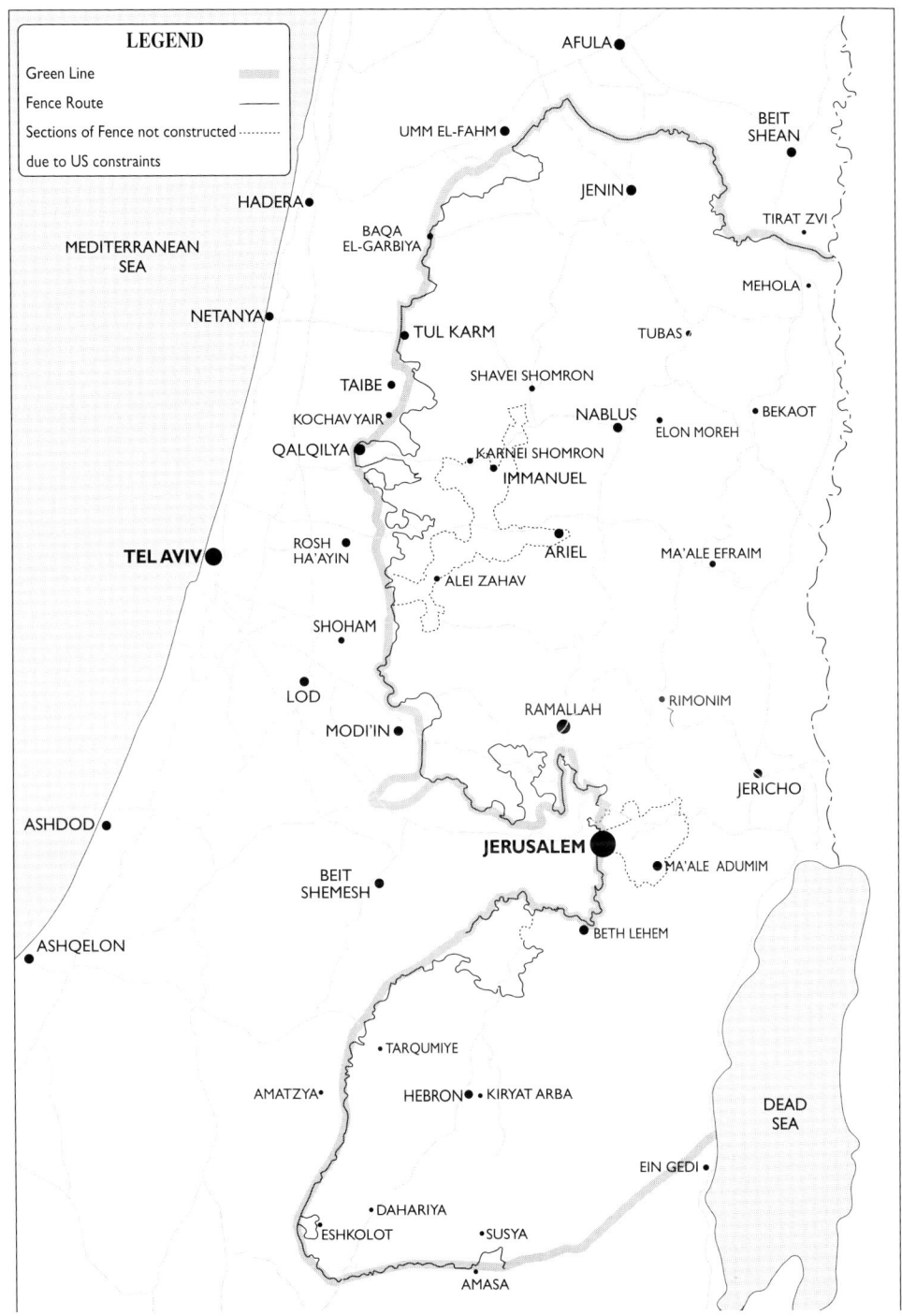

Figure 3.3 Route of the Barrier

Source: Based on the map that appeared on the Ministry of Defense website, Revised Route of the Security Fence According to the Government Decision of February 20, 2005.

tion of a barrier as a means of self-defense is legitimate as long as it is erected within Israeli borders and not on Palestinian soil. The Palestinians attempted to prevent the construction of the barrier not only through political means, such as petitioning various foreign governments and the International Court of Justice, but through violent demonstrations as well.

The construction of the barrier must be seen in the context of Israel's decision to implement the disengagement plan: on the one hand, it reflected the desire of the majority of Israel's population to separate from the Palestinians; and on the other hand, the very creation of a physical barrier created a reality of separation to which the Israeli public was gradually growing accustomed.

Prime Minister Ariel Sharon's disengagement plan was first announced in December 2003. It called for the total evacuation of all twenty-one Israeli settlements in the Gaza Strip, which are concentrated primarily in Gush Katif, but excluded the thin strip adjacent to the Israeli–Egyptian border known as the Philadelphi route. In addition, it called for the evacuation of four settlements in northern Samaria, which would facilitate Palestinian territorial contiguity in the northern West Bank as a whole.

This decision, the first time that the Israeli government decided to evacuate settlements in the Land if Israel, i.e., historical Palestine, serves as a critical precedent. The fact that a government controlled by right-wing parties legislated the move renders it even more significant. Moreover, the decision was made as a unilateral step, without demanding anything from the Palestinians in return. The disengagement plan appears to have resulted from Prime Minister Sharon's realization that Israel's continued control of a large Palestinian population threatened the existence of Israel as a Jewish and democratic state. The timing of the decision apparently coincided with increasing pressure on Sharon to embark on some initiative to bring about a profound change in the crisis in Israeli–Palestinian relations. These pressures stemmed from the fleeting episode of Abu Mazen's first government, the brief ceasefire and its rapid disintegration, expressions of dissatisfaction within the Israeli public on the lack of political movement, and the positive response received by the Geneva accords, a document published during the same period that presented a model for an Israeli–Palestinian final status agreement.

The disengagement plan enjoyed wide public support from the outset. Supporters included those interested in reaching a settlement with the Palestinians based on an Israeli withdrawal from the Palestinian territories, as well as those who doubted Israel's ability to reach a settlement with the Palestinians, whether because they did not regard the Palestinians as presenting a viable partner for reaching a settlement or because they did not think it was possible to reach an acceptable agreement with the Palestinians. The latter group therefore would prefer to determine Israel's border with the Palestinians unilaterally. Prime Minister Sharon belonged to the second group. In several interviews, he stressed that he initiated the disengagement plan because of his belief that if he did not put forward an Israeli initiative, he would come under great pressure to reach an agreement with the Palestinians that was unacceptable to him like the Geneva accords.

The widespread support enjoyed by the disengagement plan resulted to a large extent from the growing recognition in Israeli society that Israel's continuing occupation of Palestinians threatens the existence of Israel as a Jewish and democratic state, due to the demographic reality between the Jordan River and the Mediterranean Sea.

If the Jewish and Palestinian populations are not partitioned into two separate political frameworks, Israel will have to choose between its Jewish character, as the state of the Jewish people, and its democratic political system, which will not allow it to continue controlling a large Palestinian minority that is possibly on the way to becoming a majority. There was nothing new in this demographic information. However, the violence of the second intifada drove home its significance in Israeli public consciousness by making it clear that control over the Palestinians and the reality in which the two populations live and interact with one another is destructive to Israel.

A large number of political obstacles stood in Sharon's way of implementing the disengagement plan, largely because support for the plan was greater among circles outside Sharon's government than within his Likud party and his governing coalition (which largely comprised parties from the Israeli right wing). Opposition to the plan was clearly reflected in May 2004, when Sharon was pressured into holding a referendum among members of his own party, a referendum that he lost. Problems within his coalition forced Sharon to change the constitution of his government by adding the Labor Party and the ultra-orthodox United Torah Judaism party to replace the right-wing parties and the politically middle-of-the-road Shinui party that resigned from the coalition. Shinui, however, continued supporting the disengagement plan from outside the government. By mid-2005, Sharon had managed to overcome all political obstacles to the disengagement plan within the government and within the Knesset, leaving only the extra-parliamentary arena open to opponents. For their part, opponents continued their attempts to torpedo the plan by initiating demonstrations and making preparations for violent resistance to the evacuation of settlements.

According to the plan's timetable, settlement evacuation would begin on August 15 and last for a number of weeks. Israeli forces would leave the Gaza Strip a few days after the evacuation was completed. Approaching the disengagement, however, a number of questions regarding the plan's implementation remained unanswered. The original government decision of June 2004 called for the destruction or relocation of all residential structures and synagogues, and for transferring only the economic infrastructure to the Palestinians. However, this issue remained under reconsideration since the destruction of homes would have a negative impact on Israel's image abroad. It was also likely to cause operational problems due to the need to keep troops in the area after the settlers were evacuated.

Although the plan is unilateral, Israel hoped to coordinate its implementation with the Palestinians. Yet despite this desire, which grew since the death of Arafat, the level of coordination between the two parties in the disengagement remained ill-defined. Israel's political leaders also wanted the withdrawal from Gaza to be a full withdrawal, in order to enjoy international recognition of the end of Israel's occupation of the Gaza Strip. For this reason, they were eager to withdraw from the Philadelphi route as well, notwithstanding the Israeli military's staunch opposition to such a move until a solution was found for preventing the smuggling of weapons from the Sinai Peninsula into the Gaza Strip. Talks between Israel and Egypt were aimed at persuading Egypt to uphold its responsibility for preventing such smuggling. However, Egypt contended that it was unable to do so without deploying suitable military forces along the border with the Gaza Strip. This possibility, in turn, sparked bitter political debate in Israel due to concerns that it would precipitate the disintegration of the military appendix to

Israel's peace treaty with Egypt, which stipulates that a large portion of the Sinai Peninsula is to remain demilitarized. On the eve of the disengagement, the question of a suitable arrangement that facilitates the withdrawal of Israeli troops from the Philadelphi route and the subsequent establishment of a seaport and an airport in the Gaza Strip remained unresolved issues.

Complexity of the actual disengagement implementation was linked to the question of how much resistance would be staged by Gaza and northern Samaria settlers and their supporters in an attempt to thwart the plan's execution. It also depended on whether or not the disengagement would take place in peaceful conditions or under Palestinian fire. This question was important for three reasons: first, because it had potential – albeit limited – to halt the implementation of disengagement; second, because it might incur harsh Israeli reactions that would lead to the reoccupation of Palestinian cities in the Gaza Strip; and third, because the actual trauma caused by disengagement will influence the attitudes of Israeli political actors and the Israeli public regarding steps to be taken after disengagement.

➤ THE POST-ARAFAT ERA

The November 2004 death of Palestinian Authority head and PLO leader Yasir Arafat accelerated political developments within the PA, including incipient trends that began to emerge while Arafat was still alive. The transfer of power itself proceeded in an orderly fashion. This disproved concerns that anarchy might erupt in the PA, exhausted and suffering from a serious inability to function after years of armed struggle and after the disappearance of its dominant ruling figure. These concerns also resulted from the fact that during his life, Arafat himself labored to ensure that he was completely identified with and inseparable from the Palestinian people and the Palestinian revolution. However, immediately following his death, the Palestinians invoked the laws pertaining to the position of prime minister within the PA. At first Abu Mazen was appointed the head of the PLO Executive Committee, while the speaker of the Legislative Assembly served as the acting Palestinian prime minister. In January 2005, Palestinians went to the polls to elect Abu Mazen as the new prime minister. He received 62 percent of the vote, thus acquiring legitimacy for his status and direction as the new Palestinian leader.

The election results represented a profound shift in the dominant paradigm of the PA. It was clear from the support of Abu Mazen that the majority of Palestinians in the territories wanted to stop the violent confrontation with Israel that began in September 2000, referred to in internal Palestinian debates as "the militarized intifada," and instead advance the national interests of the Palestinian people through political means by negotiating with Israel. The elections also reflected efforts by the cadres of Fatah, the ruling party representing the Palestinian national-secular sector, to rebuild their party around its new leader. Fatah's status was greatly damaged during the years of the intifada for a number of reasons:

- Led by Fatah, the PLO was Israel's partner in the Oslo process. Upon the collapse of the peace process and the yielding of results antithetical to Palestinian aspirations, Fatah's association with the process not only punctured its status but destroyed its political platform. The vacuum that remained was filled by opposition and Islamic parties, primarily Hamas, which was now able to claim that it had been correct in opposing the Oslo process.

- The pressure on the PA during the intifada, in conjunction with Arafat's ruling style of building up different power centers and playing them off one another, resulted in the party's disintegration.

- The decision of Tanzim, Fatah's militia, to join the violent struggle against Israel in the form of al-Aqsa Martyrs Brigades led to the establishment of local armed gangs in different locations. Under no authority, these gangs were often seen by criminal elements as an opportunity to gain legitimacy and improve their status within Palestinian society. With their participation in the intifada, members of these groups could parade as those who sacrificed for the Palestinian people and were therefore entitled to a special status reflective of their elevated position during the intifada years, when in practice they functioned as the unauthorized local rulers of their cities, towns, and villages.

- The damage done to Fatah during the second intifada further advanced the decline of the party and the PA as a whole, which had already started to lose support due to the corrupt image of Palestinian officials. This state of affairs stemmed from the improper system of rule instituted by Arafat that enabled him to exercise complete control over the PA, in part by allowing him to grant financial benefits to close associates. The PA was thus dominated by cartels and monopolies, especially with regard to imports into the Palestinian territories controlled by people close to PA officials and their families. Arafat also prevented the establishment of an effective justice system, which could have placed restraints on the PA "robber barons."

As a result of these developments, the PA came under pressure to implement reforms even before Arafat's death. In March 2003, internal pressures combined with pressures exerted by the international community resulted in the appointment of Abu Mazen and his first government. Until then, there had been no separation between the positions of prime minister and head of the PA, both of which were held by Arafat. One important step taken by the Abu Mazen government was the appointment of highly respected Palestinian economist Salam Fayyad to the post of finance minister. Indeed, Fayyad succeeded in implementing significant reforms within the finances of the PA, promising increased transparency and accountability. However, Arafat's undermining of Abu Mazen's government and resistance to the unification of PA security services under government authority resulted in the rapid disintegration of the new government. For its part, Israel was unwilling to take steps to help or strengthen Abu Mazen, and the Israeli–Palestinian ceasefire collapsed. And despite any of the implemented reforms, the years of the intifada only exacerbated the PA's image of corrupt leadership because while the general population's standard of living plummeted during the period, the general impression was that regime officials were not sharing in their poverty. Rather, they maintained their relatively high standard of living.

The group of Fatah members who organized themselves around Abu Mazen and contributed significantly to his election was constituted by members of Fatah's younger generation, who hailed from the Palestinian territories and had substantial bases of local support. They saw themselves as engaged in a struggle with the older generation, which came to the territories from Tunisia with Arafat at the beginning of the Oslo process. From their point of view, if this group of leaders was not deposed, the Palestinian leadership would continue to suffer from its corrupt image and the party would lose its support to the Islamic opposition, led by Hamas. As such, the young generation saw itself as in the midst of an existential struggle that it must win. For its part, Hamas was successful in filling the vacuum left by Fatah because its leaders enjoyed an untarnished image of dedicating their lives to the good of the Palestinian people. This image stemmed from the group's focus on the armed struggle, the social services they provided for the population, and the fact that so far they were uninvolved in administering the PA and its institutions.

All of these trends were reflected in the two rounds of elections for local authorities held in the PA following the Palestinian presidential elections. Although Hamas did not win an absolute majority in the local elections due to the large number of independent candidates elected, its showing was better than Fatah's. Clearly the determining factor in the local elections was the image of each candidate as either corrupt or clean.

The next stage of the PA election process was to be elections for the Palestinian Legislative Council, the Palestinian parliament, originally scheduled for July 17, 2005 but postponed, probably to January 2006. The results of these elections will to a great extent determine the character and policy of the PA and the entity's relationship with Israel. It is difficult to assess which factors will have the greatest influence on Palestinian voters: basic support of Abu Mazen's policies and his desire for continued quiet on the security front and renewing the political process, in the hope that this will result in improvements to economic and social living conditions, or the degree to which candidates are seen as either corrupt or honest. If the determining factor is the former, people will vote for Fatah. If the determining factor is the latter, they will vote for Hamas.

These developments were likely to make Hamas an even more important political actor than it was before, which meant that the changes underway within Hamas were extremely significant. Hamas has generally operated according to the model of Hizbollah in Lebanon, aspiring to enter the political arena while keeping its armed militia intact. However, entering the political arena demands a more acute sensitivity to the mood of the Palestinian public and the ability to adapt its positions accordingly. In this context, there may be initial signs of moderation of the group's position toward Israel. For the first time, Hamas officials made statements indicating a willingness to join the Palestinian consensus in favor of the establishment of a Palestinian state alongside the state of Israel, within the borders of the West Bank and Gaza Strip that were occupied in 1967. The group attempted to resolve the contradiction between this position and its overall ideology by expressing willingness to accept an armistice lasting many years, if Israel agreed to withdraw to the 1967 borders.

Abu Mazen realized that for him, a prerequisite for success was the cessation of violence. If the violent struggle with Israel were renewed, he would be weakened and public support for Hamas would increase. For this reason, his regime's first major efforts were aimed at reaching a ceasefire and persuading two principal parties to agree:

al-Aqsa Martyrs Brigades, associated with Fatah, and Hamas. In many ways, the task of convincing al-Aqsa Martyrs Brigades was greater, because the Brigades were made up of smaller armed groups without a clear organizational framework and under no central authority, with very few actual ties between them. In contrast, Hamas was a well-ordered organization with armed groups that usually operated under the authority of its political leadership. Thus, in the case of Hamas, it was enough to convince the political leaders. In the case of the Brigades, each individual group had to be persuaded.

Largely due to his particular personal style – which avoided confrontation and snuffed out any suggestion of a possible civil war as anathema to the Palestinian national ethos – and out of his political weakness, Abu Mazen's efforts rested on persuasion. It was doubtful if the PA could emerge victorious from a confrontation with these armed groups. In fact, from the outset, some local leaders of al-Aqsa Martyrs Brigades demonstrated to Abu Mazen his own weakness by striking at and humiliating local representatives of the Palestinian regime. In the meantime, however, Abu Mazen was successful in persuading Hamas to agree to a ceasefire, despite the group's official opposition to a ceasefire and its lack of faith that Israel would honor a ceasefire.

To facilitate Hamas's retreat from its extreme position, *tahdiya*, "calming," replaced the previously-used term *hudna* to refer to the ceasefire, which suggested that any lull in fighting was less of a formal, organized commitment to an end of the violence. Nonetheless, the inference is that Hamas decided to join the ceasefire after realizing it suited the mood of the Palestinian street, and that as a political movement it could not afford to ignore its source of power in Palestinian society. The opportunity to join the Palestinian political process from a position of strength also appealed to Hamas.

Although the other groups who participated in the fighting with Israel likewise joined the ceasefire, it remained incomplete and fragile. For one reason, it was not based on a bilateral agreement determining what each side could and could not do. Rather, each side had its own interpretation of the agreement. Israel, for instance, held that it could continue arresting people suspected of involvement in terrorist activity, though such arrests sometimes resulted in the death of suspects who attempted to escape. In a development of that sort, various Palestinian groups saw themselves as free to respond, for instance by firing mortar shells on Israeli settlements. Moreover, the problem of al-Aqsa Martyrs Brigades was not solved. Members of these armed groups refused to relinquish their military power and forfeit their independence. As a result, most violations of the ceasefire were actually carried out by groups related to Fatah – not Hamas. The ceasefire also became hostage to Palestinian domestic struggles, and when Hamas was unhappy with Abu Mazen's decision to postpone the elections it took the opportunity to initiate intensive exchanges of fire with Israel.

The issue of suspects "wanted" by Israel was a special problem. As long as they saw themselves as pursued by Israeli forces, they had no interest in maintaining the ceasefire. Israel gave the PA a list of wanted people whom it was willing to stop pursuing, on the condition that they surrendered their arms, promised to stop their violent activities, and remained under Palestinian supervision. The PA had a hard time implementing this agreement, and therefore at least initially it was applied only in limited instances. Moreover, a number of friction-producing issues still remained unsettled, the most prominent of which concerned the Philadelphi route, where

Palestinian efforts to smuggle weapons into the Gaza Strip continued. Israeli efforts to prevent weapons smuggling generally led to Palestinian casualties and violent Palestinian responses.

Maintaining the ceasefire depended on Abu Mazen's ability to reform the Palestinian Authority's security services and develop their capabilities. Abu Mazen himself repeatedly emphasized his intention to actualize the principle of the regime's monopoly over the use of force. The aim of the reforms was to limit the number of Palestinian security services to three – a civilian police force, a force attuned to domestic security, and a force for external security – and to place them all under the unified supervision of the Palestinian interior minister. In the six months after Abu Mazen's election, the reforms were implemented at a snail's pace due to the opposition of centers of power and individuals with vested interests. The struggle surrounding this issue highlighted Abu Mazen's weakness stemming from his desire to avoid confrontation.

Building up the capabilities of the Palestinian security services was largely dependent on the aid of external entities in cooperation with Israel. While outside parties including the United States, the European Union, Egypt, and Jordan expressed great willingness in this area, Israel's position was split. On the one hand, officials understood the need to build up Palestinian security capabilities. On the other hand, they did not trust the Palestinians and feared the quick collapse of the ceasefire, and therefore objected to arming the Palestinians and giving them control over territory. This is also why the transfer of cities to Palestinian control, decided on at the Sharm el-Sheikh summit in February 2005, proceeded so slowly. By early July, only two cities were transferred to Palestinian control: Tulkarm and Jericho.

Despite the weakness of the ceasefire and the slow pace of security reforms, the ceasefire demonstrated remarkable resilience. Many incidents that under other circumstances might have resulted in its total collapse turned out differently, with each party making do with only limited responses and stopping before crossing the line of no return. In the first six months after Abu Mazen's elections, the only large-scale attacks were two suicide bombings, one in Tel Aviv at the beginning of the ceasefire; and one in Netanya in July 2005, which then prompted an exchange of fire in the Gaza Strip. Palestinian and international reactions to the attacks forced the Islamic Jihad, responsible for the bombings, to conceal and deny its involvement and stopped the consequent exchanges of fire. In other incidents, the killing of terrorists during Israeli arrest attempts resulted in reprisals such as the launching of mortar shells against Israeli settlements and localities. Nonetheless, the cycle of reprisals and counter-reprisals was arrested quickly and remained circumscribed. Hamas's success in the local elections also resulted in the group's own assessment that it had more to gain by maneuvering along a political path. Specifically, Hamas was anxious to avoid a situation in which Israel failed to withdraw from the Gaza Strip due to the deterioration of the security situation, as Palestinian public opinion would undoubtedly attribute this to the group's actions. These facts increased the ceasefire's chance of survival.

The first months of Abu Mazen's presidency yielded a positive balance sheet. He succeeded in establishing public legitimacy for his regime, initiating a democratization process that included the Islamic opposition, and taking the first steps of reform within the PA in general and the Palestinian security services in particular. However, groups supporting him began to sense that he was losing his momentum because he did not take determined steps to remove the old guard and officials from Tunisia or purge Fatah of corruption. From their perspective, one expression of Abu Mazen's weakness and loss of momentum was his inability to overcome the opposition of Palestinian prime minister Abu Ala. A dynamic of competition evolved between the two, in which Abu Ala sabotaged Abu Mazen's attempts to implement reforms in the Palestinian Authority as well as his policy of agreeing to enter a dialogue with Israel on coordination of the disengagement plan.

➤ THIRD PARTIES AND THEIR INFLUENCE ON ISRAELI–PALESTINIAN RELATIONS

Prime Minister Sharon's disengagement plan enjoyed widespread support in the international community because it appeared to be an opportunity to end the deadlock. Equally significant was that most international actors attributed great importance to beginning the process of evacuating settlements and saw it as a significant precedent. The Bush administration's initial reaction to Sharon's announcement of the disengagement plan was one of confusion, as Sharon had not first consulted with the American government. However, the Americans quickly regained their composure and decided to support the plan.

Prime Minister Sharon initiated a running dialogue with the American administration through his close advisor Dov Weisglass, using this dialogue to persuade the Americans to take steps to help him overcome the political obstacles within Israel. The most prominent of these steps was the exchange of letters between Prime Minister Sharon and President Bush on April 14, 2004, in which the United States expressed its support of the disengagement plan and extended certain guarantees to Israel. A section of Bush's letter that was interpreted as supporting Israel's annexation of settlement blocs within the framework of an eventual final status agreement attracted special attention, although the text itself is actually phrased much more cautiously: "In light of new realities on the ground, including already existing major Israeli population centers, it is unrealistic to expect that the outcome of final status negotiations will be a full and complete return to the armistice lines of 1949."[1] Much less attention was paid to the Israeli obligations enumerated in Sharon's letter, in which the prime minister promised to limit construction in the settlements and to evacuate all illegal outposts.

Once President Bush emerged victorious from the American elections of November 2004 and launched his second term in office, it remained to be seen whether there would be any changes in American policy in the Middle East and, in this context, in American policy *vis-à-vis* the Israeli–Palestinian conflict. Personnel changes among top US leadership, the most important being the replacement of Secretary of State Colin Powell with National Security Advisor Condoleezza Rice, were likely to impact on this question, although the actual effect was still unclear. On the one hand, Rice played a major role in designing the first administration's policy toward the Middle East and supported this policy much more than Powell. On this basis, continuity in American policy could be expected. On the other hand, Rice stood to be a stronger and more influential secretary of state than was Powell, due to her close relationship with the president. Through her, State Department officials, many of whom displayed discomfort with America's Middle East policy during the first Bush administration, might have greater influence over American policy on the Israeli–Palestinian issue. The new situation in light of Arafat's death and the disengagement plan was also likely to influence American policy. The United States had an interest in supporting Abu Mazen and the democratization processes within the PA. Abu Mazen's agenda served America's primary interest in the Middle East and helped America improve its image in the Arab Middle East.

The United States displayed early signs of increased willingness to become more involved in Israeli–Palestinian relations. One development was the appointment of two officials to coordinate aid to the Palestinians: General William Ward, deputy commander of American ground forces in the European command, appointed as the military coordinator of the Palestinian issue in February 2005; and James Wolfensohn, the former World Bank president who was named as the economic coordinator. Both were appointed by the American secretary of state but understood their positions as coordinators for the Quartet (the United States, the European Union, Russia, and the UN secretary-general), and were therefore surrounded by an international staff. General Ward also tried to expand his responsibilities by attempting to serve as a coordinator between Israel and the Palestinians in the realm of security, which failed due to Israeli opposition. A second sign of increasing American willingness to get involved was the secretary of state's visit to Israel and the Palestinian Authority in February 2005. A third sign was the warm reception enjoyed by Abu Mazen during his visit to Washington in May 2005. The United States was likely to play an important role following implementation of the disengagement plan, due to the large gap between the Israeli and the Palestinian positions on various issues and the distrust that dominated the relations between the parties. The United States was assumed to have the ability to bridge some of these gaps and prevent crises that were likely to develop as a result.

The European Union initially opposed the disengagement plan based on suspicions that the plan's real aim was to fortify Israel's hold over the West Bank. However, the Europeans came to the conclusion that the disengagement plan was preferable to a stalemate, and that the important precedent of settlement evacuation could generate momentum in the desired direction of settling the conflict as a whole. The US and European Union support for the disengagement enabled the Quartet to pass a resolution supporting the plan, on the condition that it be the first step in implementation of the roadmap. The death of Arafat and the election of Abu Mazen reinforced the Europeans' sense that a window of opportunity had opened and that it should be taken advantage of in order to restart the political process. In this context, British Prime Minister Tony Blair convened a conference in London in early March 2005. The conference was meant as a show of support for Abu Mazen and the rejuvenation of the political process, and concluded with commitments by European leaders and leaders of other countries to provide the PA with aid.

The positions of Arab countries also underwent similar evolution with regard to the disengagement plan. They initially adopted the Palestinian perspective, which viewed the plan suspiciously as a tool by which Israel intended to strengthen its grasp on the West Bank – a plan the Palestinians referred to as "Gaza first and last." However, as time passed, the leading Arab countries reached the conclusion that implementation of the plan was better than a deadlock. They also came to regard it as representing potential for a change in Israel's approach and creating a process that could result in Israel's withdrawal from the occupied territories.

This perspective altered Egypt's approach to the Israeli plan, and the shift was reflected in improved Egyptian–Israeli bilateral relations, marked by the return of the Egyptian ambassador to Tel Aviv; Egypt's willingness to release Azzam Azzam, who had been convicted and incarcerated in Egypt on charges of spying for Israel; and, most importantly, Egyptian efforts to contribute in any way possible to ensure the success of the disengagement plan. In this context, Egypt convened the Sharm el-Sheikh summit with Abu Mazen and Sharon and helped facilitate a dialogue between the PA and Palestinian opposition groups, which eventually resulted in a ceasefire. It worked to help the Palestinian administration implement reforms in its security services, and was also willing to provide additional assistance through consultation and training. It attempted to mediate between Israel and the PA regarding security issues. Finally, out of a desire to create the conditions for a full Israeli withdrawal from the Gaza Strip, it indicated a willingness to help solve the Philadelphi route problem by committing to stop smugglers. Its condition for this, however, was permission to station suitable border patrol forces along the border with the Gaza Strip, which would require an amendment to the military appendix to the Israeli–Egyptian peace agreement.

Due to the importance of the issue, President Mubarak assigned responsibility for the Israeli–Palestinian portfolio to Egyptian intelligence minister Omar Suleiman, considered one of the strongest members of the Egyptian regime. His intense involvement was indicated by his many visits to the PA and to Israel. Egypt's activity reflected not only the desire to end the Israeli–Palestinian conflict, but also stemmed from Egyptian concern that developments in the Gaza Strip following the Israeli pullout could damage its interests. For example, the Egyptian regime feared that Islamic elements might take control of the Gaza Strip after the disengagement, which could impact on domestic affairs within Egypt, where the ruling regime was already waging a concentrated struggle against Islamic movements that threatened its stability.

Jordan was less involved in mediating Israeli–Palestinian relations, due to the tension that has always existed between Jordan and the Palestinians. Jordan sees the Palestinians as a potential threat to its ruling regime, and the Palestinians carry with them memories of Jordanian rule in the West Bank and suspect that Jordan still wants to control them. While various Palestinian circles were willing to cooperate with Jordan, the dominant approach was rejection and caution. The Jordanians were willing to help the PA by training policemen and moving the Palestinian Liberation Army – the Badr Brigade, located in Jordan and integrated into the Jordanian military – to Palestinian control, in order to provide the Palestinians with an accessible and well-trained military force. Israel has objected to such a step, and the Palestinians themselves have also not decided whether or not to adopt the idea. Egypt and Jordan's efforts to generate an Arab consensus in favor of the process spurred by the disengagement plan – based on the Saudi initiative that was endorsed in 2002 at the Beirut Arab League Summit – failed because of the divide characteristic of the Arab collective.

➤ POST-DISENGAGEMENT

After Prime Minister Sharon successfully overcame the many political pitfalls that threatened implementation of the disengagement plan, the main political actors assumed that the plan would in fact be carried out. They therefore began to focus on "the day after" by examining the possible scenarios that could develop following the Israeli withdrawal and by taking steps to encourage developments in the directions they favored.

The most visible phenomenon in this context was the divergence of Israeli and Palestinian positions. Prime Minister Sharon and his spokespeople created the impression that Israel would make do with the withdrawal from the Gaza Strip and northern Samaria for the time being, and that Israel did not presently see a need for additional steps, certainly not for the resumption of talks with the Palestinians. While Sharon remained formally committed to the roadmap, his commitment has included a large number of constraining factors. First, it was argued, the Palestinians must fulfill all of their commitments during the first phase of the roadmap, including the elimination of the terrorist infrastructure, before Israel would take any steps of its own. Second, Sharon repeatedly expressed his position that there was no chance of reaching a final status agreement with the Palestinians, and that the most he was willing to accept is a long-term provisional arrangement with them. In other words, the Israeli position appeared to be that the process should stop indefinitely with the implementation of the second phase of the roadmap, which stipulates the establishment of a Palestinian state with temporary borders. It remained unclear whether Sharon intended reaching this interim arrangement through unilateral steps or through an agreement with the Palestinians.

For his part, PA leader Abu Mazen clung to his position of calling for the immediate resumption of final status agreement negotiations and for reaching a settlement as soon as possible. Abu Mazen, who was the force behind the Palestinian group that participated in the Geneva initiative, appeared to believe that it might be possible to reach an agreement with Israel along the lines of the Geneva accords.

Internal developments in both the Palestinian and Israeli camps were also likely to have a significant influence on the development of relations between the two parties. The electoral success of Hamas and the weakening of Fatah were liable to hinder dialogue between the two sides and enhance Israel's tendency to adopt unilateral steps. In an extreme scenario, developments could even result in the removal of Abu Mazen from the Palestinian political arena. In the Israeli camp, indications were that partners in the government coalition – the Likud and the Labor parties – would disagree on important issues shortly after the disengagement, which in turn would likely cause the disintegration of the coalition, the toppling of the government, and early elections. This outcome was even more probable because both parties had an interest in sharpening distinctions between their opinions in preparation for elections that, regardless of developments, were not far off. Early elections would make it more difficult for Israel to make decisions in the coming months.

Socioeconomic developments in the Palestinian camp would also prove important. After more than four years of the intifada, the Palestinian economy was in dire straits, unemployment was high, and poverty was widespread. Internal changes in the PA, the relative cessation of violence, and the disengagement plan have created expectations for rapid improvement throughout the Palestinian public. Yet despite the willingness of many international parties to help the Palestinians, it would be difficult to bring about a quick visible change in the living conditions of Palestinians. The fragile nature of the ceasefire limited Israel in granting Palestinians more freedom of movement, easier passage out of the Palestinian territories, and the ability to work in Israel. This in turn meant that quick improvement of the Palestinian economy was extremely difficult to achieve. The absence of rapid change might cause disappointment throughout the Palestinian public, and this was likely to have political ramifications, if not violent ones.

International players were also expected to have an influence on developments after disengagement. The nature and extent of their impact would depend on their degree of involvement, the scope and urgency of their assistance to the Palestinians, and their ability to influence both parties to agree on a compromise-based process. American policy and activity were considered the key to effective handling of international involvement in this realm.

Finally, as in other instances in the past, violent confrontations that might erupt between the two sides, due to diverging positions and growing frustration and disappointment, stood to dictate developments and prevent additional progress on both the unilateral level and bilateral levels.

Note

1   See President Bush's letter to Prime Minister Sharon, April 14, 2004, <http://www.knesset.gov.il/process/docs/DisengageSharon_letters_eng.htm>.

CHAPTER ➤➤➤

4

Approaching a Nuclear Iran: The Challenge for Arms Control

EMILY B. LANDAU

As in the previous few years, Iran's ultimate aim in the nuclear realm has still not been fully defined. Iran continues to deny that it has ambitions in the military realm, but evidence first revealed in mid-2002 has grown and strengthened suspicions that it does. The fact that Iran maintained a secret nuclear program for close to twenty years, coupled with the widely held view that Iran did not need the program for energy purposes, convinced many that it was likely pursuing a military option.

Assuming Iran does have military ambitions, it is not clear whether attempts to persuade or force it to reverse course will be successful, or whether Iran will finally become a nuclear weapons state. What is clear in light of the past three years' trial of Iran and its nuclear program, however, is that the relative complacency with which many states regarded the supposed effectiveness of non-proliferation efforts in the nuclear realm – primarily through the instrument of the Nuclear Non-Proliferation Treaty (NPT) – is no longer warranted. The NPT has been exposed as ill-equipped to stop a determined proliferator, and attempts to continue to deal with Iran on the basis of its NPT commitments are proving highly problematic.

Yet perhaps due to the specific challenge that the prospect of a nuclear Iran poses to the international community, the thrust of international efforts over the past two and a half years has not focused on the limitations of the non-proliferation approach. Rather, emphasis has rested on preventing the particular scenario of a nuclear Iran from materializing. Since the August 2002 announcement of an exiled Iranian opposition group that it had evidence of Iran's clandestine nuclear program, international actors have attempted to form a clearer picture of the activities in Iran, with the aim of ensuring that Iran remains in compliance with its commitment according to the NPT not to develop nuclear weapons.[1] This effort to make sure Iran upholds its NPT commitments has taken precedence over attempts to address the apparent limitations of the NPT regime itself, and the failed NPT Review Conference of May 2005 was a major setback for meaningful discussion of the treaty's weaknesses, especially as far as having any relevance for dealing with Iran.

Much of the commentary on Iran, therefore, has tended to sidestep serious consideration of the challenge that Iran poses to the current approach to non-proliferation. Instead, numerous articles written in this period recount the actions and policy directions of the three major players that have surfaced in the effort to restrain Iran: the IAEA, the US, and the EU-3 (France, Germany, and Great Britain), to examine whether they have influenced Iran's basic willingness to cooperate. Analysts have assessed and evaluated different mixes of incentives and coercive measures (carrots and sticks) that the EU-3 and the US have considered and presented to Iran as a means of convincing/compelling it to adhere to its NPT commitments and verify that its course is indeed, as Iran claims, non-military.

In contrast, the present discussion seeks to focus directly on the implications for arms control by analyzing current efforts to bring Iran back to the fold in terms of the logic of non-proliferation that finds expression in the NPT. The problematic effort to ensure that Iran is complying with its NPT obligations will then be considered in relation to a different strategy for controlling the dangers associated with nuclear proliferation that could very well gain prominence down the road, especially if Iran becomes a nuclear state. The question is whether a different kind of arms control, which focuses on stabilizing relations among states, can play a positive role in mitigating the dangers of nuclear weapons in such a scenario.

➤ CONFRONTING IRAN: THE ARMS CONTROL DIMENSION

Generally speaking, arms control efforts and initiatives can assume different forms, including (but not limited to) disarmament, non-proliferation, nuclear weapons-free zones (NWFZs), arms limitations agreements, demilitarized zones, and confidence and security building measures (CSBMs). In terms of arms control logic, a distinction exists between agreements that seek primarily to eliminate weapons (disarmament and non-proliferation), and measures (such as demilitarized zones and CSBMs) that attempt, by enhancing inter-state stability, to decrease the chance that states might employ dangerous weapons. The dominant strategy for dealing with Iran over the past three years has been diplomacy aimed at ensuring that Iran upholds its NPT commitment not to become a nuclear state. Significantly, these international efforts, while carried out as diplomacy and negotiations, have continued to adhere to the logic of non-proliferation: namely, making sure that Iran maintains the non-nuclear status it accepted.

Two problems arise in this regard. The first is that if Iran is in fact attempting to become a nuclear state, the non-proliferation approach embedded in the NPT has already proven to be of limited utility for dealing with it. Notably, the success of the NPT depends to a large degree on states' acceptance of the idea that nuclear weapons are an inherent cause of insecurity in the international arena. However, while many states may adhere to this principle, the NPT at the same time implicitly recognizes that a nuclear deterrent capability can also enhance state security. This is why the NPT compensates states that agree to make the concession in security terms that is implied by remaining in a non-nuclear status. The potential security benefits tied to possession of nuclear weapons – which non-nuclear states are aware of and which might motivate them to develop or acquire these weapons – are not, however, addressed directly by the treaty. Rather, the treaty basically assumes that non-nuclear states will accept that it is in their interest to remain non-nuclear. The NPT's lack of explicit attention to genuine security concerns and possible interests to go nuclear (beyond its exit clause) hampers its ability to prevent a determined proliferator from working towards a military nuclear capability.[2] In this sense, the case of Iran, more than it weakens the NPT, starkly exposes the deficiencies already inherent in the NPT and the logic of non-proliferation, which focuses on weapons without reference to interests and inter-state relations.

The second and related issue is that confronting Iran's behavior has turned into a political process in which self-appointed strong states have taken the lead. The problem is that these actors have continued to carry out this process in the name of the NPT, while at the same time their very behavior has highlighted the constraints involved thereby. Thus, while the EU-3 and the US (as well as the IAEA) have ostensibly been focused on the objective, even technical issue of determining whether evidence of non-compliance exists or not, in fact, clear-cut objective criteria have been very hard to define and agree upon, let alone nail down in this process. Much has hinged on different interpretations of the evidence, analyses of what constitutes a breach of commitments and what should be done about it, and assessments of the degree of cooperation and goodwill that Iran was considered to have displayed in this process. Needless to say, at times the gaps in interpretation work in Iran's favor. But in any case, the fact that dealing with Iran is a political, rather than objective/technical process further highlights the conundrum of approaching nuclear proliferation as if it were merely a matter of denying capabilities by ensuring compliance with the NPT. In fact, political interests are involved at all levels and need to be recognized as part of the arms control approach.

➤ THE DIPLOMATIC ROUTE: THE EU-3 AND THE US

The major diplomatic effort toward Iran over the past two years has been driven by the EU-3. International pressure on Iran increased in 2003, following the conclusion of IAEA Director-General ElBaradei in June of that year that Iran had failed to meet its obligations under its Safeguards Agreement. This led the IAEA Board of Governors in September to issue an ultimatum to Iran, calling on it to cooperate fully with the IAEA, suspend all further uranium enrichment-related activities, and join the Additional Protocol by October 31.[3] Europe began to consolidate its own policy toward Iran, and ten days before the ultimatum was to expire, the EU-3 states were able to reach a deal with Iran, whereby Iran committed itself to the terms of the ultimatum in a bilateral (EU-3–Iran) agreement. However, this deal broke down eight months later (in June 2004), with Iran's announcement that it was resuming activities related to uranium enrichment, claiming that Europe had not upheld its end of the 2003 bargain, which Iran understood as a European commitment to remove the case of Iran from the IAEA Board of Governors agenda by June 2004.[4] This was a clear indication that the seemingly technical issue had become a political one.

In the months that followed, the EU-3 continued its efforts to strike a new deal with Iran in order to reinstate the suspension on activities related to uranium enrichment. These efforts finally resulted in an agreement signed in Paris in mid-November 2004, just prior to an IAEA meeting that otherwise might have seen Europe acquiesce to the

US preference to refer Iran to the UN Security Council for possible sanctions. The EU-3, however, was determined to continue on its diplomatic route and not allow the crisis to escalate to the point of UN referral. Under the terms of the new agreement, Iran once again committed itself to full cooperation and transparency with the IAEA, continued voluntary implementation of the Additional Protocol pending ratification, and suspension of all enrichment and reprocessing activities. The suspension was to be sustained while negotiations with the EU-3 proceeded on a mutually acceptable agreement on long-term arrangements relating to Iran's nuclear activities. The long-term agreement is to provide objective guarantees that Iran's nuclear program is for peaceful purposes only; it is also intended to provide guarantees on nuclear, technological, and economic cooperation and firm commitments on security issues.[5]

It was hoped that this deal would overcome the shortcomings of the initial agreement from 2003; nevertheless, the question of whether Iran's suspension of uranium enrichment activities was to be temporary (Iran's understanding) or long-term (Europe's interpretation) was still not finally resolved. Iran viewed its commitment to suspension as a confidence-building measure directed to the international community, but in no way as a negation of its legal right (according to the terms of the NPT) to continue with this activity after a period of time. In May 2005, on the eve of the 2005 NPT Review Conference, Iran expressed dissatisfaction with the pace of negotiations with Europe, and once again stated its intention to resume activities related to the enrichment process at the Isfahan Uranium Conversion Facility.[6] The lack of agreement between Iran and the EU-3 over what was meant by "suspension" was reflected in their different reactions to the possibility that Iran may resume enrichment activities. Whereas the EU-3 stated that it would view resumption of uranium processing as a clear violation of the November agreement, Iranian Foreign Ministry spokesman Hamid Reza Asefi was quoted as saying that Iran's "decision to resume part of our activities in Isfahan is fully compatible with the essence of the Paris agreement."[7]

Over the course of the past two years, the US adopted a noticeably less active stance *vis-à-vis* Iran. It has basically taken a back seat to European diplomatic efforts and accepted the deals made, while raising the "stick" of sanctions and/or military action from time to time to remind Iran of the possible consequences if it doesn't fully cooperate with international demands not to pursue activities of a military nature. But while the US has serious doubts whether Iran can be trusted, the option of referring the case of Iran to the UN Security Council has not been pursued as forcefully as it might. The EU-3, for its part, has pressed for more active American backing and endorsement of its negotiations with Iran in order to increase their prospects of success. Following Bush's European tour in early 2005, an understanding was reached whereby the US would demonstrate more support for the EU-3's efforts to present incentives to Iran, and the EU-3 in return would back the US demand to refer the case to the Security Council if it became clear that Iran was not cooperating. This was expressed in early March by Secretary of State Condoleezza Rice, who noted US willingness to lift its objection to an Iranian application to join the WTO, and also to allow the sale of spare parts for Iranian commercial aircraft.[8] The US was signaling a stronger commitment to economic incentives for Iran.

While diplomatic efforts continued during the first half of 2005, their record was not encouraging, and it remained far from clear whether the international community was significantly closer to bringing Iran into true cooperation. Iran's May 2005 state-

ments regarding its intention to renew enrichment activities were just a further negative development in a process rife with setbacks. In fact, what has emerged from the
diplomatic efforts so far is that for a number of reasons Iran has enjoyed a distinct edge
in the negotiating process. First of all, Iran has had the advantage of being focused on
a single interest (to resume enrichment activity, most likely to advance a military
program), and it has been highly determined in this regard. The relevant international
actors dealing with Iran, on the other hand, have not been coordinated in their
approach, and have been influenced by other sets of interests,[9] although of late they
have made some attempts to present a more coordinated approach.

In addition, Iran has conducted the negotiations in a manner that maximizes its
advantage over the international community: it has continued to cooperate to the
degree necessary to ward off harsh measures, while pushing its program forward all
the while. By not taking any action too provocative that might catapult the international community into more concerted action, the dynamics of the negotiations have
played to Iran's advantage. This has worked especially well because the IAEA and the
EU-3 became highly committed to the success of the diplomatic route – especially as
an alternative to the possible US use of force. They have capitalized on any indication
of cooperation on the part of Iran as justification for continuing the negotiations.
While the negotiating process has slowed down Iran's progress, it has nevertheless
(albeit unintentionally) enabled Iran to move its nuclear program forward.

Iran's position has been enhanced by the fact that almost everything in this process
revolves around interpretation. Missing is a set of precise criteria for non-compliance
that would trigger automatic referral to the Security Council, short perhaps of a
"smoking gun" such as a nuclear test. Determining whether Iran was cooperating with
the international community in the negotiations process has proven most difficult to
define in clear operational terms – for example, with regard to the understanding
reached between the US and the EU-3: what in fact will serve as a decisive indication
that Iran is no longer cooperating, or that diplomacy has failed? So far, every instance
of non-cooperation has merely served as the basis for a new round of diplomatic efforts
to secure cooperation. Indeed, Iran and the EU differed over whether an Iranian decision to renew enrichment activities was a violation of the November 2004 Paris
agreement. But even with a firm European stance that this was in fact a violation, it
was not certain that this would automatically lead to a referral of the case to the
Security Council, as one might have thought in light of the US–EU-3 understanding.
While the EU-3 reacted to the Iranian intention to resume activities with a harsh
warning as to the "consequences" that Iran would face, the net result was more negotiations. Iran indicated its willingness to negotiate with the EU-3 before making its
final decision on this issue.[10]

Meanwhile, in May 2005 there were hints that the EU might finally be losing
patience with Iran's negotiating tactics, but also indications that European negotia-
tors sought to put off any confrontation with Iran until after the June presidential
elections, in the hope that the victor would be willing to accept incentives in return for
abandoning the problematic aspects of Iran's nuclear program.[11] Iran likewise hinted
that its patience was wearing thin, and that if prevented from using nuclear technology
for civilian applications, it might begin to lose respect for the NPT.[12]

In early June, Iran finally agreed to postpone plans for enrichment until late July in
order to allow the EU-3 time to prepare an acceptable proposal.[13] Some ten days later,

in yet another twist to its nuclear narrative, Iran admitted to having conducted small-scale experiments in plutonium separation for five years after 1993, the date it had asserted as the completion of all such experiments.[14] This once again highlighted the issue of the secret nature of Iran's nuclear program. Nor did the victory in the June elections of the conservative Mahmoud Ahmadinejad give cause to believe that there would be a change in Iran's position on the nuclear issue. If anything, fears grew that Iran will be less willing to make concessions.

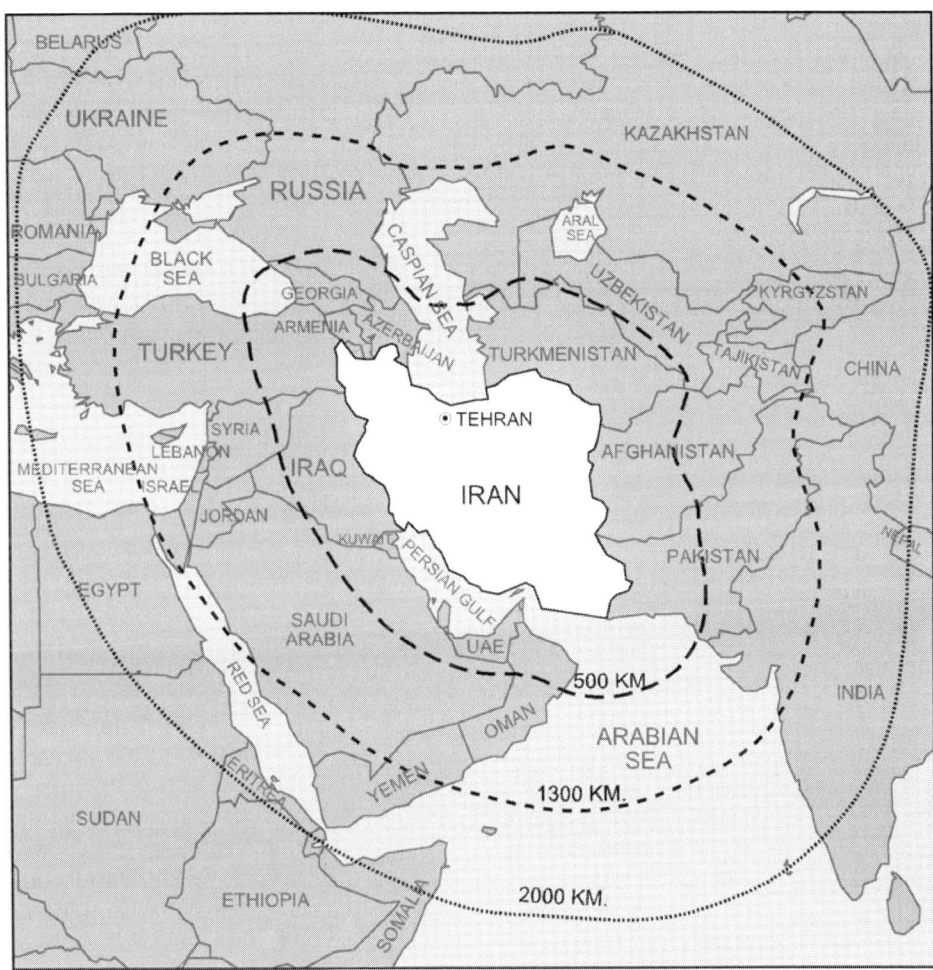

Figure 4.1   Range of Iran's Ballistic Missiles

Thus, the process so far has seen persistent EU-3 efforts to convince Iran through diplomacy to back down from the weapons option, thereby rejecting the use of force to realize this goal. In arms control terms, however, these efforts are still rooted in the problematic logic of non-proliferation. Complicating matters is the fact that the political process has been conducted as if it were indeed simply a matter of exposing evidence of non-compliance, while the actual dynamics have been more complex.

Moreover, the combination of incentives/threats that Iran views as attractive/fearful enough to cause it to rethink its interests or reverse its course has yet to emerge. And even if an agreement were reached whereby Iran cooperates in a manner that the international powers consider to be credible, unless Iran's motivation to go nuclear is addressed there is no guarantee that it won't find a way to continue its efforts.

Cognizant of the inherent limitation in the logic of non-proliferation efforts that do not take state security interests into account, as well as the related problems that constrain the current diplomatic efforts, the world must recognize that the option of a nuclear Iran is a viable possibility.[15] This is especially the case as time plays to Iran's advantage – the more time that goes by, the closer Iran comes to a military option. The question then becomes what are the options for mitigating the potential dangers, should this scenario materialize.

Clearly, the ramifications of a nuclear Iran go far beyond the question that will be considered below. If Iran becomes a nuclear state, this would impact upon all states in the Middle East and have important implications for Egypt, Saudi Arabia, Turkey, and others. It would have major significance in terms of the Gulf sub-regional context and dynamics (figure 4.1). How would other states relate to a nuclear Iran? What role would the issue of possible misperceptions and miscalculations play in regional inter-state relations? Would this be an impetus for states to consider new coalitions, or increase their own motivation to proliferate? For Israel, there are different options regarding a nuclear Iran and implications for Israel's policy of nuclear ambiguity. As far as arms control and non-proliferation are concerned, there would be serious questions to pose as to the future of the NPT, as well as whether in the final analysis it is perhaps simply easier for the international community to deal with a nuclear Iran than to try to arrest its proliferation. All of these issues and others are relevant to the possibility of a nuclear Iran, but the following discussion is limited to the question of deterrence between Israel and Iran.

➤ DETERRENCE: A BASIS FOR NEW RULES OF THE GAME?

To begin to explore how to mitigate the dangers inherent in the option of a nuclear Iran, we can consider the other dynamic parallel to diplomatic efforts to control Iran: periodic threats of attack exchanged between Iran and Israel (and/or the US) over the nuclear issue. So far, this dynamic, which has a history dating back to the 1990s, has been more on the sidelines, with the major developments taking place in the context of the diplomatic negotiations. But if Iran becomes a nuclear state, this nascent dialogue of deterrence between Israel and Iran could very well become a more central dynamic.

Specifically, the mutual deterrent messages between Israel and Iran are significant in light of the fact that current international efforts to stop Iran could very well end in failure. In that case, the major issue would become how to stabilize Israel's deterrent relationship with a nuclear Iran, and how to better deal with the possible incentives to proliferate among additional states in the Middle East. Both of these issues are likely to increase the motivation to control the dangers of proliferation by attempting to strengthen inter-state stability, as a first stage in a regional security dialogue and process. Thus, it is important to consider the nature of the threats issued at the present stage and assess whether and how mutual deterrence between Israel and Iran might be stabilized down the road.

➢ Israel

On the very possibility of Iran becoming a nuclear state, Israeli positions have been unequivocal as far as the severity of this outcome. On a number of occasions high-level officials have noted that they view this as an existential threat to Israel,[16] and Israel therefore is making every effort to convince the international community that Iran must be stopped. However, statements are much less forthright (or more ambiguous) when it comes to Israel's response to the emerging threat.

As to the question of whether Israel might carry out a preemptive strike against Iran's facilities, beyond a few statements that have hinted at this possibility, most prefer to emphasize that Israel has "answers," while leaving the operational definition of this in the realm of ambiguity. There have been cases where preemptive intentions have been attributed to Israel by non-Israeli sources, in light of Israeli statements on the severity of the pending threat, or simply due to the fact that Israel already carried out a preemptive strike against Iraq in 1981. In some instances Iran has reacted to such publications as if Israel itself made the statements. The appointment of former Air Force Commander Dan Halutz as the next IDF Chief of Staff in early 2005 also raised some speculation as to whether there was any connection between this decision and possible plans to strike Iran's nuclear facilities.[17]

Israel has demonstrated sensitivity to the possible escalatory effects of communicating through threats, and it seems that a conscious effort has been made not to provoke Iran. Thus, emphasis has been placed on Israel's need to strengthen its defensive capabilities in the case of an attack by Iran, but not on preemptive action. An interesting case regards statements made by Defense Minister Shaul Mofaz in the context of a program in Farsi on Voice of Israel Radio, in which Mofaz answered questions from listeners in Iran. The article in *Ha'aretz* that reported on this radio exchange noted in its opening sentence that Mofaz had intensified Israeli warnings to Iran, and for the first time confirmed that Israel is considering action to destroy Iran's nuclear capability.[18] However, the possibility of preemptive action was inconsistent with another Mofaz statement quoted in the same article that emphasized Israel's defensive/deterrent position toward Iran: Mofaz was quoted as saying that Israel has no hostility toward the Iranian people and no plans for attacking Iran. However, if Israel were to be attacked, then it would use all means to defend its territory and people.

This inconsistency can be explained by other reports that the article in *Ha'aretz* was actually based on a mistranslation of what Mofaz said on the program. Accordingly, the statement that seemed to indicate preemptive plans was taken out of context – Mofaz was in fact relating to the possibility that Iran would obtain nuclear weapons and threaten to use them against Israel. Confirmation for the view that all Mofaz's statements were made in a defensive mode, and that his overall message was characterized by restraint, was provided by the Voice of Israel Farsi Section Director, Menashe Amir.[19] This incident is significant because the hint of preemptive plans on the part of Israel led Iran to react with severe threats of its own, highlighting the potential danger of such rhetoric. The attempt by the Israeli foreign ministry to set the record straight in turn underscored Israel's desire to avoid escalation and use of threatening language.

A similar case of Israel's sensitivity to unintended escalation regards statements attributed to Air Force Commander Eliezer Shkedi in late February 2005. An Associated Press report from February 21 quoted Shkedi as saying that "Israel must be prepared for an air strike on Iran in light of its nuclear activity." The Israeli press reported the next day that Shkedi had been misquoted, that the IDF was most upset over this incident, and that Israeli officials feared that the mistaken attribution would be viewed as provocation against Iran.[20]

➤ Iran

Iran's threats to attack have been directed against Israel and the US, with the aim of deterring them from carrying out a surprise preemptive strike against its nuclear facilities. Since the 2002 revelations, Iran has clarified the degree of ruin and destruction that would follow a strike on its installations. For example, a report on Iran's preparations for a surprise attack on its nuclear facilities and other strategic targets was published in late March 2005 by *Al-Hayat* of London. According to the report, in such a scenario, "the objective is to deliver a harsh blow to the US and its ally Israel at the outset, and then to expand the arena, in light of international efforts to contain the crisis and limit its scope and intensity, so as to ignite the whole region."[21]

While much has been made of the Iranian Shehab-3 missiles that carry slogans about their ability to destroy the State of Israel, as well as a few extremely hostile official statements – often taken as a signal of the existential threat that a nuclear Iran would pose – Iranian statements of the past two years have primarily been in reaction to fears of attack. They have tended to come in the wake of possible hints that Israel (or the US) was preparing for a preemptive strike. This was the case, for example, in August 2003 when a *Washington Post* article that expressed US concerns that Israel might be considering the option of attacking Bushehr[22] prompted an Iranian warning to Israel that such an attack would be a serious mistake on Israel's part and exact a heavy price. Following Mofaz's statements in December 2003, Iranian Defense Minister Ali Shamkhani was quoted as saying that Iran would strike Israel with all weapons at its disposal if Israel ventured to attack. Again, this reaction to what it was thought that Mofaz said illustrates not only that Iranian threats came mainly for deterrence purposes, in the wake of fears that Israel was planning an attack, but also how easily states can be caught up in a dynamic of unintended escalation when issuing deterrent threats in the nuclear realm.

Overall, it appeared that the threats on both sides were basically of a deterrent nature, designed to ward off attack rather than initiate it. Moreover, attempts were made, at least by Israel, to avoid escalatory rhetoric that could provoke Iran. While one could argue that there might be other reasons for this pattern – such as that Iran has still not achieved a nuclear capability that would enable it to attack Israel, or that Israel prefers to remain ambiguous about its intent in order to increase the element of surprise if it did decide to attack – nevertheless, indications are that neither side wished to risk the actual use of nuclear weapons, and both were trying to communicate this to each other through initial deterrent exchanges.

➤ THE POSSIBILITY OF STABLE DETERRENCE

The question of whether a stable nuclear deterrent relationship can develop between Israel and Iran is as yet an open one, and full analysis of this issue is beyond the scope of this discussion. Nevertheless, some initial insights can be advanced. Those that support the thesis that stable deterrence between Israel and Iran would be extremely difficult to achieve have hinged their argument primarily on their view of Iran as a radical, aggressive, and revisionist state. This, together with problems of communication between the two states, which in themselves create fertile ground for misperceptions, has rendered stable deterrence a most problematic option in this arena.[23]

However, the fact that it will be difficult to create a stable deterrent relationship and new rules of the game for the Middle East with a nuclear Iran doesn't mean that both sides won't have an interest to begin working in this direction. Assuming that Israel and Iran have a mutual interest to avoid nuclear war – and from the initial deterrence dialogue so far, it seems that such a mutual interest exists – this is the common ground that can serve as the starting point for both sides to work toward enhancing stability, and ultimately for beginning to create new security structures in the Middle East.

The absence of communication channels between the two sides, and the fears of misperceptions and miscalculations, would be the very motivation for both to try to create these links. The agreement upon initial confidence-building measures (CBMs) in the US–Soviet sphere in the 1960s and early 1970s – especially the Hot Line established between the two in the wake of the Cuban Missile Crisis of 1962 – is most instructive in this regard. Moreover, as to the characterization of Iran as a hostile, radical, and revisionist state that would never stabilize its relations with Israel, it is worth recalling assessments of the Soviet Union advanced in the US in the early years of the Cold War. According to the April 1950 US National Security Council (NSC) memorandum, "the Soviet Union, unlike previous aspirants to hegemony, is animated by a new fanatic faith, anti-thetical to our own, and seeks to impose its absolute authority over the rest of the world."[24] The Soviet Union, from the late 1940s to the mid-1950s, was viewed as revisionist, fanatical, and bent upon destroying the integrity of the US. Over the subsequent years, this harsh image was modified in light of developments in the bilateral sphere.

The mutual interest to stabilize the deterrence equation, and to deal with the possible motivation of additional states in the Middle East to develop or acquire nuclear weapons, could very well sharpen the logic of creating a framework for regional security dialogue in the Middle East. In such a regional dialogue, the focus needs to be security interests and threat perceptions of the participating states, in order to alter the context within which WMD exist. In the first stage, this would involve steps to improve communication among regional states and negotiation of initial CSBMs.

Interestingly enough, the strongest official statements that have been made in support for renewing regional security dialogue in the Middle East have come from the Director-General of the IAEA, Dr. Mohamed ElBaradei. While it might seem that ElBaradei is the one person who would have the strongest interest to address arms control in the nuclear realm solely in terms of the NPT, perhaps he is uniquely positioned to realize just how limited the NPT is for dealing with actual cases of proliferation. Over the past few years, ElBaradei has made the case for dealing with the sources of insecurity that motivate states to attempt to obtain WMD.[25] His strong support for the recent European initiatives with regard to Iran has lauded the drive to place all issues on the table. In accordance with his view that the nuclear program is merely "the tip of an iceberg," the head of the IAEA has welcomed dealing with nuclear ambitions in the context of inter-state relationships.[26] With this in mind, he has advocated holding a strategic dialogue on creating a WMD-free zone in the Middle East and made an effort to convene such a regional forum in early 2005. He has so far been unsuccessful in this endeavor, due to an inability to secure agreement on an agenda for the meeting.

To be sure, the argument that stable mutual deterrence cannot simply be assumed is well taken. Nonetheless, the nascent dialogue of deterrence moves the dynamics of arms control in the direction of focusing on the relevant states themselves: their threat perceptions, security interests, and inter-state relationships. While the option of a nuclear Iran is clearly a highly undesirable outcome, this might be what pushes states in the Middle East to the point where they begin to seriously consider their security relations in the regional sphere.

The diplomatic process that the world has seen so far is still intimately bound up with the logic of non-proliferation as expressed in the NPT. The EU-3 and US are caught in a situation where they are seemingly adhering to objective and technical criteria, whereas in reality this is not the case. The prevailing view that the NPT is the relevant frame of reference for these discussions has effectively constrained the ability of states to influence Iran's calculations. The party that has gained from this situation is Iran, which has skillfully used the "interpretation gaps" in the NPT, as well as the EU-3 proclivity to pursue the diplomatic route, in order to play for time.

Ultimately, arms control cannot be detached from the political context of inter-state relations or treated as if it exists solely in the realm of technical arrangements. While various steps can be envisioned to enhance the ability of the NPT to deal with real cases of suspected proliferation, such as improved means of verification or tighter controls on technology, there will always be limits to technical solutions that don't take state security concerns and other interests into account. As with the case of Iran *vis-à-vis* the NPT, if interests and relations are ignored, they will come back to haunt the arms control treaties and arrangements that ignored them.

Notes

1   The major non-proliferation treaty, the NPT, seeks to prevent the spread of nuclear weapons to additional states, beyond the five nuclear states (i.e., those that tested nuclear weapons before 1967). As a non-nuclear state party to this treaty, the demand from Iran is that it take steps to reassure the international community that it is complying with its commitment not to attempt to develop or acquire nuclear weapons.

2   Emily B. Landau, "The NPT and Nuclear Proliferation: Matching Expectations to Current Realities," *Strategic Assessment* 6, no. 4 (2004): 32–36.

3   See IAEA Director-General, "Implementation of the NPT Safeguards Agreement in the Islamic Republic of Iran," June 6, 2003 (GOV/2003/40); and IAEA Board of Governors, "Implementation of the NPT Safeguards Agreement in the Islamic Republic of Iran," September 12, 2003, (GOV/2003/69).

4   Emily B. Landau and Ram Erez, "WMD Proliferation Trends and Strategies of Control,"

in Shai Feldman and Yiftah Shapir (eds.), *The Middle East Strategic Balance 2003–2004* (Brighton & Portland: Sussex Academic Press, 2004), pp. 62–67; and Gary Samore, *Meeting Iran's Nuclear Challenge,* Paper no. 21, Weapons of Mass Destruction Commission, October 2004.

5   "Communication dated 26 November 2004 received from the permanent Representatives of France, Germany, the Islamic Republic of Iran and the United Kingdom concerning the agreement signed in Paris on 15 November 2004," IAEA Information Circular, INFCIRC/637, November 26, 2004.

6   See "Iran to Resume Some Nuclear Activities," *Reuters,* May 3, 2005; and "Iran to Resume Some Nuclear Activities 'in Days,'" *SpaceWar,* May 9, 2005.

7   Parisa Hafezi, "Iran Prepares to Resume Sensitive Nuclear Work," *Reuters,* May 9, 2005.

8   Rice said: "The Europeans are engaged in a diplomacy that we support, we are now actively supporting that diplomacy with ... this decision on these two elements, and we will now see whether the Iranians are really serious about living up to their international obligations ... I am confident that if the Iranians fail to live up to their obligations and that ... if this track does not work, that there will be a referral to the [U.N.] Security Council." Rice added that the US decision put the Iranians on notice that they would not be able to split the Europeans from the Americans on this issue. See "US to Lift Objections Against Iranian Bid to Join the WTO," US Department of State, Office of the Spokesman, March 11, 2005.

9   For a concise take on the economic interests that preclude Europe from taking a stronger stand on Iran, see Thomas Friedman, "Brussels Sprouts," *New York Times*, May 11, 2005.

10 Moreover, European Union foreign policy chief Javier Solana was quoted as saying that if talks break down between Iran and the EU over Iran's resumption of enrichment activities, the file should go first to Vienna, namely, the IAEA, before the Security Council. See "Iran Nuclear Case to be Referred to IAEA if Talks with EU Fail," *SpaceWar,* May 10, 2005.

11 *New York Times,* May 18, 2005.

12 Modher Amin, "Analysis: Iran's Patience Run Out?" *SpaceWar,* May 17, 2005.

13 *New York Times,* June 5, 2005.

14 It did so when confronted with the results of laboratory tests; see *New York Times*, June 16, 2005.

15 Though not within the scope of this discussion, widely analyzed have also been problems inherent in military action as a means of stopping Iran. See, for example, Ephraim Kam, "Curbing the Iranian Nuclear Threat: The Military Option," *Strategic Assessment* 7, no. 3 (2004): 1–8.

16 For example, Meir Dagan, head of the Mossad, at the Knesset Foreign Affairs and Defense Committee in November 2003: "Our estimate is that the Iranians will continue to develop the military projects of their nuclear industry. Such a measure of warfare in their hands would for the first time pose an existential threat to Israel." *Ha'aretz*, November 18, 2003.

17 *SpaceWar,* February 23, 2005.

18 *Ha'aretz,* December 21, 2003.

19 See Ellis Shuman, "Mistranslation Leads to Heated Iranian Threats," *IsraelInsider*, December 25, 2003; and explanations in this vein published by the Israeli Foreign Ministry: "Statement on Interview by DM Mofaz on Israel Radio," Israel Ministry of Foreign Affairs, December 24, 2003.

20 "Israel Media: IAF Commander Misquoted Regarding Possible Air Strike Against Iran," China FBIS Report, February 22, 2005.

21 Quoted in A. Savyon, MEMRI, no. 218, April 7, 2005.

22 *Ha'aretz,* August 18, 2003.

23 See, for example, Gerald M. Steinberg, "Deterrence Instability: Hizballah's Fuse to Iran's Bomb," *Jerusalem Viewpoints,* no. 529, April 2005.

24 See *NSC no 68: United States Objectives and Programs for National Security,* 14 April 1950.

As to fundamental Soviet aspirations, it is further noted in this report that "the United States . . . is the principal enemy whose integrity and vitality must be subverted or destroyed by one means or another if the Kremlin is to achieve its fundamental design."

25 For ElBaradei statements, see interview with *Al-Arabiya*, February 25, 2004 (*Federal News Service*, 25 February 2004); Mohamed ElBaradei, "In Search of Security: Finding an Alternative to Nuclear Deterrence" CISAC, Stanford University, November 4, 2004; interview with *Le Monde,* March 23, 2005 (FBIS translated text); interview with *Arms Control Today,* March 2005; and interview with *Al-Sharq al-Awsat*, May 6, 2005 (FBIS translated text).

26 ElBaradei differentiates what he refers to as the "political" approach of the EU-3, from the "technical" approach of the IAEA, while in fact, interestingly enough, he is advocating a different logic of arms control – one that focuses on states and relationships. See interview with *Arms Control Today*, March 2005.

CHAPTER ➤➤➤

5

International Terror, with Iraq at its Hub

YORAM SCHWEITZER

In mid-2005, the international campaign against terrorism sparked by al-Qaeda's September 11 attack in the United States approached the end of its fourth year. Its success – or its failure – has remained an open issue. The counterattack began in October 2001 with a massive conventional military strike in Afghanistan launched by an American-led international coalition. It continued a year and a half later with a war waged by the United States and its allies against the regime of Saddam Hussein. The official goals of the war in Iraq included toppling Saddam Hussein's regime due to its support of terrorism and its alleged possession of non-conventional weapons, which included the concern that the regime was liable to distribute these weapons to terrorist organizations. Furthermore, the war in Iraq would also send a message to other recalcitrant regimes. The American-led action signaled that if they pursued the same path, their fate would be the same as that of Saddam Hussein and the Taliban regime, which had facilitated the expansion and dissemination of terrorism from Afghanistan to other places around the world.

Although the military campaigns in Afghanistan and Iraq both resulted in the fall of the regimes, the overall campaign against international terrorism has not defeated the main forces within al-Qaeda and its affiliates that fuel the terrorism. Moreover, while the linkage between the proliferation of international terrorism and the war in Afghanistan was clear to everyone, the linkage between terrorism and the war in Iraq raised many eyebrows, even before the war began. Doubts as to the merits of the war arose in many countries, including a number of prominent American allies in the war on international terrorism. France and Germany, for example, did not regard Iraq as an important actor in the operations of al-Qaeda and its affiliates. They questioned the wisdom of a military strike against Iraq and expressed concern regarding the possible unexpected results of aggressive action, as well the action's regional implications in general and its contribution to the war on international terrorism in particular.

More than two years after the war in Iraq, charges resound that the war not only failed to strike at and wipe out the instigators of international terrorism, but instead sparked an increase in terrorism. The contention is based on an assessment of events since the end of the war, particularly over the past year, during which terrorism has played a major role in destabilizing Iraq and has added to the already existing difficulties of reconstructing the country. Terrorism has claimed the lives of many Iraqi and foreign citizens working towards reconstruction. According to critics, the war created a pretext for anyone aiming to encourage the recruitment of a new generation of volunteers into the ranks of the international jihad movement, a movement that had been in decline following heavy losses sustained in the war in Afghanistan. The chaotic situation and massive loss of life in Iraq has facilitated the dissemination of the radical Islamic argument that foreign forces are attacking Muslims and occupying Iraq. This challenge, they argue, necessitates mobilization for a war to defend both an Arab country with a great history and Islam from those who wish to humiliate and destroy them. According to the dedicated jihad activists, the attackers include the Americans, their allies who took part in the wars in Afghanistan and Iraq, and the Arab regimes that supported them.

As a result, Iraq has come to constitute a magnet for young Muslims from around the Arab world, particularly Saudis, Syrians, and Kuwaitis. These volunteers came to the country to fight alongside local Iraqis against foreign forces, in effect casting the arena as a Muslim battlefield resisting an invasion of Crusaders.[1] This view provides a basis for claiming that all Muslims are obligated to join the jihad not only out of community obligation (*fard al-kifaya*) but out of personal obligation (*fard al-`ayn*). This, in turn, provides the ideological–theological basis for the obligation of *al-istishhad* (self-sacrifice in defense of Islam), which underlies the campaign of suicide attacks underway in Iraq over the past few years.[2] It also underlies the recent massive campaign of kidnappings of foreign and Arab hostages, which has aimed partly at political blackmail and fundraising through ransom, and partly at instilling fear in foreigners residing in Iraq for practical and propaganda purposes. Many kidnappings have ended with showcase executions of foreign hostages, generally citizens of leading coalition countries and their Arab supporters. The perpetrators of these showcase executions were apparently primarily men in the ranks of Abu Mus`ab al-Zarqawi, a Jordanian, and his Iraqi allies.

Yet while the major effort of al-Qaeda and its affiliates has focused on Iraq during the past year, these organizations have continued operating against Arab regimes outside of Iraq as well, primarily in Saudi Arabia, Egypt, and the Gulf states. They have also been active in Europe – at this stage, primarily in Britain and Russia – as well as the Caucasus. The July 2005 attacks in London, which were preceded by attacks in Madrid and Istanbul, might very well foreshadow the spread of terrorism from Iraq into other regions, including the capitals of Western countries.

➤ **THE IRAQI ARENA**

Since its onset in March 2003, the campaign of suicide attacks has gradually transformed Iraq into the most prominent theater of suicide terrorism. The hundreds of suicide attacks that have occurred in Iraq in a relatively short time have added a new quantitative dimension to the phenomenon that is unprecedented in scope. Over the past two and a half years, between 350 and 500 suicide attacks were carried out in Iraq (figure 5.1),[3] resulting in the death of more than 3000 people. Noteworthy also is the lack of clarity surrounding the nationality, organizational affiliation, and exact number of the suicide terrorists, apparently stemming from the chaotic state of the Iraqi security forces and the coalition forces' lack of comprehensive, well-organized intelligence information. There are varying assessments regarding the percentage of local suicide attackers relative to the percentage of "foreigners" in the ranks of the international jihad. There may also be an intentional lack of clarity in the dissemination of accurate information to the public. In addition, the small percentage of suicide attackers who have been arrested after failing to carry out their mission or having second thoughts reinforces the dearth of information, as the interrogation of such operatives could shed additional light on this complex subject.

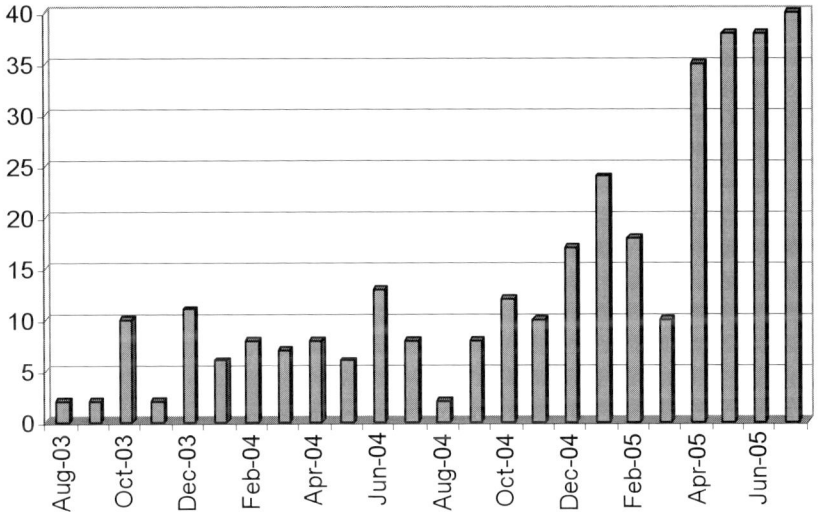

Figure 5.1   Number of Suicide Attacks in Iraq by Month, August 2003–July 2005

Despite the uncertainty regarding their identity, based on the claims of responsibility issued, it appears that much responsibility for casualties and property and infrastructure damages lies with the members of the international jihad groups and networks led by the Jordanian Abu al-Khalayilah, better known as Abu Mus'ab al-Zarqawi. Indeed, most of the targets in Iraq appear on the list of Zarqawi's primary enemies targeted for attack, included in the letter meant for Osama Bin Laden and seized among the possessions of Zarqawi's chief assistant, who was apprehended in Iraq in February 2004. In this letter, Zarqawi mentioned the need to strike at the foreign forces that had arrived in Iraq, most importantly the Americans and their allies, the Shiite Muslims – whom he referred to as snakes and as worse than the foreign infidels – and Iraqi institutions and the local security forces who were collaborating with the enemy. The strategic concept that Zarqawi advanced focused on the mass killing of Shiites, aimed at provoking them to undertake violent reprisals against the Sunnis. This, in turn, would create a state of chaos and regime instability, as well as damage to the security forces, institutions, and senior administration officials. The final result would be to prevent the establishment of a stable government and to terrorize foreigners into halting reconstruction efforts and leaving Iraq.[4]

In an unusual step, Bin Laden agreed to formally name Zarqawi – who belonged to the camp of activists who identified with Bin Laden but who had neither sworn allegiance to him nor were under his command – as the commander of al-Qaeda in Mesopotamia and his representative in the region. This appointment came in response to Zarqawi's proposal that he accept Bin Laden's authority and swear his allegiance to him, and that Bin Laden in turn recognize him as his prince. Thus, Zarqawi's strategy of warfare was officially adopted as the overall strategy of warfare of al-Qaeda and its affiliates in the campaign in Iraq.[5] Overall, there is no doubt that holy suicide, as employed by al-Qaeda and its affiliates, is a commercial symbol and a unifying organizational value[6] and has become their primary weapon in the war in Iraq. Furthermore, it has enabled them to deliberately transform Iraq into the new jihad arena and their main locus of warfare against the enemies of Islam.

Holding hostages has been an important component in the terrorists' strategy of warfare and intimidation, and alongside the suicide attacks, the period 2004–5 has witnessed the proliferation of kidnappings: of foreign civilians, diplomats, businesspeople, media correspondents, and managers and employees of foreign companies. Those kidnapped were of many nationalities, including American, British, Italian, French, Swedish, Turkish, Macedonian, Korean, Japanese, Chinese, Nepalese, Egyptian, Jordanian, Kenyan, and others. The total number of people kidnapped was approximately 150, including a few dozen Westerners.[7]

A number of groups have taken responsibility for the kidnappings, most with some Islamic identification, such as the Islamic Army of Iraq; Ansar al-Sunna; al-Tawhid w'al-Jihad; and the Salafist Brigades of Abu-Bakr al-Sidiq. Most of these names are probably code names for groups identified with Zarqawi, but the possibility certainly exists that Iraqi criminal elements have taken advantage of the chaotic state of affairs in Iraq to reap financial profits as well. It is also possible that in some instances criminal elements received major payoffs in return for releasing hostages unharmed.

One of the clearest elements of the kidnappings has been the psychological and propaganda war waged by means of foreign hostages. The kidnappers offer mercy if the governments meet their demands and withdraw their forces from Iraq, and threaten to execute the hostages in the event they refuse. In a substantial number of instances, the kidnappers have carried out the threat of executing hostages and have taped the executions, distributing the horrifying images on videotapes and compact discs that are prominently displayed on Islamic internet websites. Clearly, the goal has been to pressure the governments of the hostages who beg for their lives and plead on video to meet the demands of their captives and withdraw troops from Iraq. In this way, the kidnappers have attempted to transfer responsibility for the death of hostages to their leaders, to deter foreigners from coming to Iraq, and to encourage new Muslim cadres to join the struggle against the West. The West, on the other hand, appears in these videotapes as weak, humiliated, and a force that despite its superior military strength can be overcome.

The hostage issue in Iraq is notably similar to the kidnapping and imprisonment of approximately 100 foreign hostages in Lebanon between 1984 and 2001. These kidnappings were carried out by Hizbollah, operating under code names, in collaboration with Iran, which offered assistance and even benefited from both financial and political profits resulting from hostage release deals. Hizbollah tortured to death a number of hostages, including Colonel William Higgins and William Buckley, and the cruelty with which Hizbollah treated its hostages was covered widely by the media. In the Iraqi case, the harshness of Zarqawi's men and their intentional provocation of the West are notably pronounced. So is their clear desire to demonstrate to all their hatred and their contempt for the lives of their enemies, which complement their eagerness to sacrifice their own lives. All of this is presented as part of their religious beliefs and their interpretation of Islamic commandments, which not only allow them to act in this manner toward whomever they define as an infidel – Christian, Jew, or Muslim – but obligates them to do so.

➤ TERRORISM OUTSIDE IRAQ

➢ The Arab Arena

In 2004–5, a number of prominent attacks were carried out by different al-Qaeda affiliates against Arab countries they regarded as serving the infidels. According to the worldview of these groups, treason against Islam is an even more serious offense than the sins of the Western infidels. Replacing these Arab regimes with Islamic regimes is thus regarded as a priority. Terrorist attacks against representatives of Muslim countries have targeted Egypt, Algeria, Pakistan, and Turkey, and other attacks have been conducted on these countries' own sovereign soil.

The most intensive campaign of this kind has waged in Saudi Arabia, resulting in a war between members of the Saudi branch of al-Qaeda and the Saudi royal family, which is a primary target of these attacks. The terrorism campaign in Saudi Arabia, at whose core lies suicide attacks, began in May 2003, and the most recent suicide attack occurred in Riyadh in late December 2004. The attacks, which included assassinations and the kidnapping of foreigners, targeted the oil industry and foreign presences in the country, with an overall aim of damaging the Saudi economy and intimidating foreign workers into leaving the country. They also targeted Saudi Arabia's own security forces. These attacks compelled the Saudi regime to deviate from its usual practices and to fight terrorism intensively to a degree it had never done before. Previously the Saudis were concerned about not taking actions that would impel al-Qaeda leaders and operatives working outside of the kingdom to begin operating within Saudi borders. For this reason, they had hitherto refrained from a head-on confrontation with them and had not prevented them from operating. After the wave of terrorist attacks that rocked the country, the Saudi authorities felt they had no choice but to take action against these elements, initiating a campaign that is still underway. In addition to a number of major failures during this campaign, the Saudis have achieved some notable successes, resulting in the death of a number of al-Qaeda leaders in Saudi Arabia and the arrests of many others.[8] Saudi Arabia even hosted an international conference on the war against terrorism perpetrated in the name of Islam. At this conference, regime leaders declared the need to mobilize and dedicate themselves to a war on terrorism carried out in the name of Islam by a group that represents only a small portion of all Muslims.

Egypt is another country where terrorist attacks have resumed over the past few years. During much of the 1990s, Egypt suffered from violence at the hands of Islamic forces, primarily members of al-Gama'a al-Islamiyya and the Egyptian Islamic Jihad. It appeared that the regime succeeded in neutralizing these groups' operations with a combination of repressive measures against terrorism, dialogue with some of the group leaders, which led to a ceasefire, and the expulsion of leaders of many of the other groups from the country. However, the last year and a half has witnessed a number of terrorist attacks in Egypt, focused primarily on tourist sites in the Sinai Peninsula and in Cairo. Apparently the perpetrators of these attacks were local Egyptians who identify with the concept of international jihad, although so far their links to elements outside of Egypt have not been unequivocally established. The first prominent attack was carried out in October 2004 at a number of vacation spots in Sinai, where guests included Israelis and tourists of other nationalities. Thirty-four people were killed in the attack, including twelve Israeli citizens. The attack in July 2005 in Sharm el-Sheikh, also in Sinai, took approximately ninety lives and injured more than 200. Here too the attack targeted tourist sites, which constitute a primary sector of the Egyptian economy. It is also likely that the second attack was timed to coincide with the trial of those involved in the Sinai attack of October 2004 or the anniversary of the July 1952 Free Officers Coup in Egypt, and in this way to undermine the legitimacy and stability of the current regime. Once again a connection between the attackers and international jihad elements has not been clearly established, despite the fact that the rhetoric in claiming responsibility for the attack indicates at the very least links to the movement's ideological worldview.

Two additional attacks aimed at tourists and carried out by suicide terrorists rocked Cairo during April 2005. The first took place on April 7, and was perpetrated by a suicide terrorist who detonated a belt of explosives in the Khan al-Khalili Market in Cairo, causing the deaths of two tourists, one from the United States and one from France. The second attack was carried out on April 30 by a man who opened fire on a bus of tourists in Cairo and then committed suicide by detonating an explosive device he was wearing. Furthermore, the fact that Egypt has been an important target for al-Qaeda and its affiliates was reflected in Iraq, where the Egyptian ambassador Ihab al-Sherif was kidnapped and executed in July 2005.

International jihad activists have also actualized their threats to strike at the Gulf states. In Qatar, a suicide attack was carried out by a terrorist who detonated a car bomb next to a theater and a local British school. The terrorist, Egyptian in origin, killed a British citizen and injured twelve other civilians.

➤ Great Britain

Terrorist networks belonging to the international jihad movement continued to strike at targets in European cities during 2005, this time focusing their operations on London. In July, two waves of attacks were carried out against transportation targets in the British capital. During the first wave, which took place on July 7, four suicide terrorists of British nationality and Pakistani descent detonated explosive devices they were carrying in bags on three trains and a bus. During the second wave, which occurred exactly two weeks later, an attack was again attempted on transportation targets. However, this time the explosive devices were smaller and, presumably due to a technical mishap, no one was killed and only one person sustained minor injuries. Although still early in the investigation, the London attacks appear to have been carried out by a terrorist network based in Britain that relies on an infrastructure located in various cities around the country (Leeds, Luton, and London). A number of the suicide terrorists communicated with elements outside of the country. Some of the attackers recently visited Pakistan, where they studied and apparently underwent training and perhaps even received guidance and instructions for their mission.

A Pakistani–British connection has also emerged in a number of other terrorist incidents, most of which took place outside of England. Omar Sheikh, a British citizen of Pakistani descent, was involved in the 2001 kidnapping and murder of *Wall Street Journal* correspondent Daniel Pearl. The suicide bombing at Mike's Place, a Tel Aviv bar, on April 30, 2003, was executed by two British citizens of Pakistani descent. They were apparently linked to one of the suicide terrorists involved with the July 7 attack in London, who, it turns out, visited Israel approximately two months before the attack in Tel Aviv.[9] Yet another example of the Pakistani–British connection are the arrests of senior al-Qaeda operatives in Pakistan, which were carried out during 2004 and which led to the arrest of a number of operatives in Britain, including 'Isa al-Britani and a number of other operatives from South Africa.[10]

The fact that Britain eventually became a target of al-Qaeda and its affiliate is an expected surprise of sorts. Indeed, the heads of the British security services declared that an attack in their country was only a matter of time. This assessment appears to be based not only on the intelligence information presumably at the disposal of the British security services, but also on the number of attacks carried out in other countries that originated in Britain. Examples include British citizen Richard Colvin Reid's December 2001 attempt to blow up an American Airlines flight with a shoe bomb, and the subsequent arrest of Saajid Badat's, who had planned to carry out an attack similar to Reid's. Both operations were planned by Khaled Sheikh Muhammad, the planner of the September 11 attacks and part of the al-Qaeda central command. Other examples include the involvement of Zacharia Moussaoui, who lived and operated in London, carrying out terrorist activities under the auspices of al-Qaeda in the United States, and the arrest of members of a terrorist network whose members planned attacks, including the use of the biological toxin ricin, which is considered a non-conventional warfare agent.[11] Furthermore, Britain has served as one of the European strongholds of the international jihad movement, where proponents enjoyed the liberal nature of British society and its well-known standards of civil liberties, which draw a very fine line between free speech and incitement. Countries suffering from the terrorism of al-Qaeda and its affiliates, such as Saudi Arabia and France, have been critical of Britain for just this reason.

➤ Russia and the Caucasus

In 2004–5, Russia and the Caucasus states continued to suffer from suicide attacks. The most prominent of these attacks occurred with two female suicide terrorists blowing up two Russian Air Sibir planes on August 25, 2004, causing both planes to crash and killing eighty-nine passengers and crew members. Another attack, also carried out by a female suicide terrorist, targeted a Moscow subway station, killing ten people and injuring about fifty others. An Islamic group calling itself the Islambouli Brigades claimed responsibility for these attacks.[12]

The most dramatic event with regard to the Russian–Chechen conflict took place at an elementary school in the city of Beslan in northern Ossetia. Approximately thirty-two terrorists armed with weapons, explosives, and explosive belts took control of the school on September 1, 2004, the first day of the academic year. They then demanded the release of their comrades who were imprisoned in Russia and the withdrawal of Russian troops from Chechnya. The hostage crisis lasted for three days and concluded with the death of about 340 of the hostages, half of whom were children. The vast majority of terrorists were killed and only a small number were apprehended. Statements by the perpetrators of the attack itself, and of other Chechen spokesmen as well, linked the attack to the international jihad against the infidels.[13]

The attack in Beslan was to a large degree a replay of the Chechens' previous ruthless attack in a Moscow theater in October 2002, which ended in the death of 129 people during a rescue operation by the Russian security forces. In Beslan, the terrorists enhanced their operational preparations for another possible rescue attempt by arming themselves with large quantities of weapons, explosives, and gas masks, and by using hostages as human shields against all rescue attempts. The incident finally ended in a major tragedy, with an eight-hour battle between the terrorists and Russian security forces who were aided by local armed elements.

Other suicide attacks carried out during the past year in the Caucasus included an attack in Georgia in May 2005 and another in Uzbekistan on May 13 when a male suicide bomber was shot dead near the Israeli embassy in Tashkent. These attacks must be understood as part of the proliferation of the phenomenon among local elements identifying with international jihad.

➤ THE ONGOING WAR ON TERROR

The terrorist attacks that have taken place since September 11, 2001 reflect the magnitude and complexity of the challenge facing the leaders of the campaign against the trans-border terrorism of al-Qaeda and its affiliates. Since September 11, a growing number of countries have joined the ranks of countries struck by terrorism, whether through international jihad operatives acting on their own soil or against their citizens in other countries. In Iraq, the ongoing campaign of terrorism that followed the war has increased in intensity over the past year and has resulted in a large number of casualties from among the local population, the military forces in the country, and foreign aid agencies. This campaign is not expected to subside any time soon and instead might escalate, unless its leaders, first and foremost Zarqawi, are captured or killed and a stable government system is established in Iraq. However, if the situation remains as it is at the moment, the Iraqi context is likely to continue to serve international jihad activists as a pretext for promoting its activities outside of Iraq. It will also continue to serve as an arena for international jihad activists to actualize the movement's supreme unifying principle – the willingness for self-sacrifice in the name of Allah – in order to expand its application throughout the world.

The trend of expanding the borders of international jihad suicide attacks to new countries, as in the cases of Britain, Egypt, Bahrain, and Georgia, is likely to continue.

If the concern that the "Iraqi model" might be exported to other countries actually pans out, other grandiose modes of operation that have been demonstrated over the past year, primarily in Iraq and Chechnya, are liable to become tools in the hands of international jihad terrorists in other countries as well. Such modes of operation include kidnapping, siege attacks with hostages, and blowing up planes in mid-flight. In addition to demonstrating power and constituting psychological warfare against their enemies, these types of attacks are an important strategic component for Bin Laden and his associates in their campaign to cause enormous financial damages to their enemies.

It is clear that a considerable amount of the efforts that the West invests in a campaign against radical Islamic terrorism must transcend mere intelligence gathering and operations to thwart attacks, and must focus on creating an ideological structure to serve as an alternative to the one offered by Bin Laden and others like him. This effort must be led by religiously trained Muslim leaders with high moral standing in the Arab world and must be supported by prominent Muslim leaders living in the West. Only the formulation of a common and unified position and the public support of these Muslim leaders against the narrow extremist interpretations of Islam offered by the leaders of international jihad will serve to decrease the number of young people joining the ranks of international jihad and the acts of murder these young people are carrying out in the name of Islam around the world.

Notes

1   PRISM Occasional Papers, 3 no. 1 (2005): "Arab Volunteers Killed in Iraq: An Analysis," <http://www.e-prism.org/images/PRISM_no_1_vol_3_-_Arabs_killed_in_Iraq.pdf>.
2   Yoram Schweitzer and Sari Goldstein Ferber, *Al-Qaeda and the Globalization of Suicide Terrorism*, JCSS Memorandum no. 76 (Tel Aviv: Jaffee Center for Strategic Studies at Tel Aviv University, 2005).
3   The source of the statistical data included in this article is the author's database maintained at the Jaffee Center for Strategic Studies. Statistics concerning Iraq vary, with some American sources counting more than 500 suicide bombers since March 2003.
4   Dexter Filkins, "US Says Files Seek Qaeda Aid in Iraq Conflict," <www.nytimes.com>, February 9, 2004.
5   "Bin-Laden in a Speech to the Iraqi People," at: <www.memri.org.il>, December 30, 2004.
6   *Al-Qaeda and the Globalization of Suicide Terrorism*.
7   Edward Wong, "American Is Among 4 Captives Seized in Baghdad Kidnapping," <www.nytimes.com>, November 3, 2004.
8   <www.reuters.co.il 3/7/2005>.
9   Ben Kaspit, "Military Intelligence: Al-Qaeda's Finger is not Pointed at Israel," *Maariv*, July 22, 2005.
10 <www.atimes.com>, October 19, 2004.
11 <www.news.bbc.co.uk>, January 7, 2003.
12 <www.guardian.co.uk>, September 1, 2004.
13 Middle East News line, "Russia Determines Saudi Link to Beslan," September 26, 2004.

CHAPTER ➤➤➤

6

Rising Oil Prices and the Middle East Economy

PAUL RIVLIN

The massive rise in international oil prices since 1999 has largely been due to increased demand rather than reduced supply. In 1973, the Arab states cut their oil production by five million barrels a day and as a result prices rocketed,[1] but since 1999, there has been little mixing of politics and economics. In addition, the energy intensity of production, or the amount of energy needed to manufacture a unit of output, has fallen significantly in developed economies. These differences mean that high oil prices have been less damaging to the world economy than 1973–85 prices. The recent rise in prices has also been much smaller: between June 2003 and March 2005, the real price of oil rose by 74 percent; in 1973–74, it rose by 185 percent. Nonetheless, negative effects of higher oil prices continue. World economic growth in 2005–6 is forecast at a slower rate than in 2004, in part because of higher oil prices.[2] Moreover, the costs and benefits of oil price rises have asymmetric effects: consumers are hit more than producers gain, and hence the overall impact is negative.[3] The weakening of the dollar since 2002 has mitigated the effect of higher dollar prices for some, most significantly in Europe.

The main economic result of higher oil prices on the Middle East was that in the period 1999–2005 the oil income of OPEC members in the region was an estimated $700 billion higher than in the previous seven years. There was also a substantial increase in the income of non-OPEC oil producers in the region. This did not, however, transform their economies, certainly not to the necessary degree, although it has eased financial constraints.

➤ THE RISE IN OIL PRICES

Between 1998 and 2004, oil prices virtually tripled in current dollars (table 6.1). In January 2005 the OPEC basket price per barrel averaged $40.24 and at the end of June it reached $54.26, over four times its 1998 average. In 1999–2005, oil revenues were on average $100 billion higher each year than they had been in the preceding seven years (table 6.2).

Table 6.1   OPEC Basket Price, 1999–2005 ($ billion)

1998	1999	2000	2001	2002	2003	2004	2005
12.28	17.47	27.60	23.12	24.36	28.10	36.05	54.26*

*June 27, 2005

Source: MEES and <www.opec.org/home/basket.aspx>.

Table 6.2   Middle East OPEC Members Oil Revenues, 1999–2005 ($ billion, current prices)

	1992-98 average	1999	2000	2001	2002	2003	2004	2005f
Algeria	7.2	8.3	14.2	11.7	12.4	16.5	22.6	25.3
Iran	14.5	16.1	25.4	21.4	19.4	26.1	32.5	32.3
Iraq	1.5	12.1	19.8	15.7	12.6	7.5	20.0	21.3
Kuwait	10.7	11.0	18.2	15.0	14.1	18.8	27.4	28.0
Libya	8.0	7.7	12.2	10.8	10.5	13.6	18.1	19.4
Qatar	3.3	4.8	7.8	6.9	6.9	8.8	13.5	13.8
Saudi Arabia	44.3	44.9	70.9	59.8	63.8	84.9	115.1	113.8
UAE	13.2	15	26.1	22.4	21.8	21.5	30.0	31.8
Total	**101.2**	**119.9**	**194.6**	**163.7**	**161.5**	**197.7**	**279.5**	**285.2**

f = IEA forecast

Source: OPEC, *Annual Statistical Bulletin*, various issues, and EIA, *International Petroleum Information*.

Between 1998 and 2004 demand for oil increased by 12 percent. Nearly 30 percent of the increase came from China, whose economy grew rapidly during that period. The US accounted for almost 20 percent of the increase. Other developed economies in the Organisation for Economic Co-operation and Development (OECD) accounted for only 8 percent, mainly because of the recession in most of Western Europe. Apart from China, non-OECD countries, primarily in the developing world, accounted for the balance.

Demand for oil grew three times as fast in 2004 as in 2003, while for the first time since the 1970s, OPEC was producing near full capacity. For many years the supply chain was gradually getting tighter from the exploration and production stages to those of transportation and refining. In 2003, Iraqi exports were disrupted by war and although OPEC production rose by 5 percent and Middle East OPEC production by 9 percent, increased anxiety about oil supplies helped boost prices by an average of 15 percent. Furthermore, in recent years billions of dollars were poured into speculative investments in oil futures. By 2004, the increase in demand caused a 28 percent average rise in prices.

While demand rose, supply was constrained by a lack of investment in the oil fields, refineries, pipelines, ports, and ships. This was due in part to the low prices and revenues in producing countries that prevailed in the late 1980s and most of the 1990s, which limited the ability to invest in the sector. Political instability in Venezuela and Nigeria and the 2003 Iraq War affected supplies, as did quota decisions by OPEC in 2002. OPEC's production between 1998 and 2004 rose by 7.5 percent, compared with a rise in world production of 9.8 percent. This was less than the increase in demand and the balance was met by a lowering of inventories. OPEC also produced more heavy oil, while markets needed more refined oil. The mismatch, the result of underinvestment in refining, caused prices to rise even when the increase in overall supply matched that of demand.[4]

Between 1998 and 2004, Middle East OPEC production rose by 6.3 percent. In 1998, its share of world production came to 29.2 percent, in 2002 to 25.8 percent, and in 2004 to 28.2 percent. In January 2005, OPEC production was 0.3 percent below the total production quota that it had set in November 2004. This was due to shortfalls in Indonesia and Venezuela. Production in Algeria, Kuwait, Qatar, and Saudi Arabia exceeded those countries' OPEC quotas. Another factor underlying high prices was the little spare production capacity in OPEC, with the exception of Saudi Arabia.[5]

Between 1998 and 2005, oil export revenues of the Middle East OPEC states rose by an estimated 200 percent and the population of these countries increased from a total of 144 million to 170 million, an increase of just over 17 percent. The unweighted average oil income per capita therefore rose by 151 percent.[6] Between 1999 and 2005, oil revenues of the four non-OPEC Middle East oil producers listed in table 6.3 were estimated to have risen by 110 percent. Their total population rose by almost 15 percent during those years and the unweighted average oil income per capita rose by 83 percent. Between 1999 and 2004, oil production in those countries declined, as did the quantity exported, but much higher prices resulted in higher revenues.

Table 6.3 Middle East, Non-OPEC Members' Oil Revenues, 1999–2004
($ billion, current prices)

	1999	2000	2001	2002	2003	2004e	2005e
Egypt	2.56	2.56	2.60	2.48	3.35	4.5	5.0
Oman	5.53	8.91	7.63	7.42	7.75	9.8	12.0
Syria	2.34	3.56	4.06	4.28	3.72	4.0	5.0
Yemen	2.13	3.40e	2.90e	3.08e	3.42e	3.9	4.2
Total	**12.56**	**18.42**	**17.2**	**17.33**	**18.28**	**22.3**	**26.5**

Source: UN Economic and Social Commission for West Asia, *Survey of Economic and Social Developments in the ESCWA Region, 2004–2004*, <www.escwa.lb>; EIU; data from Yemen from Republic of Yemen: 2004 Article IV Consultation – Staff Report; Public Information Notice on the Executive Board Discussion; and Statement by the Executive Director for the Republic of Yemen March 2005.

e = estimates.

Largely due to higher oil revenues, in 2003–4 national income growth in the Middle East and North Africa averaged 5.5 percent a year, compared with an annual average of 3.6 percent in the 1990s. In 2003–4, per capita growth in the Middle East was slower than in much of the rest of the developing world because of the faster population growth. Unemployment in the region fell from 14.9 percent in 2000 to 13.4 percent in

2004, the result of job creation measures in a number of countries, especially Algeria and Iran. Increases in government spending accounted for a large proportion of the growth of national income.

Table 6.4 Balance of Payments and Fiscal Developments, 1999–2005

Foreign Balance Effects	1990–2000 average	2002	2003	2004 (estimate)	2005 (projected)
Middle East Export Revenues	182	258.1	327.2	428.6	451.5
Current Account Balance	1.6	33.4	71.3	133.5	134.3
Egypt, Jordan, Morocco, Tunisia Export Revenues	31.8	42.5	49.8	56.3	60.8
Current Account Balance	-0.9	1.7	5.6	3.9	2.4
Algeria, Iran, Syria Yemen Export Revenues	40.9	63.2	82.4	109.7	114.6
Current Account Balance	3.2	12.0	18.5	35.5	34.9
Bahrain, Kuwait, Oman, Saudi Arabia, UAE Export Revenues	109.2	152.4	195.0	262.6	276.1
Current Account Balance	-0.7	19.7	47.2	94.1	96.4
Fiscal Effects					
Middle East Fiscal Expenditures	130.2	253.1	299.9	368.4	379.1
Fiscal Revenues	147.2	257.2	276.9	318.4	353.8
Egypt, Jordan, Morocco, Tunisia Fiscal Expenditures	33.6	35.1	38.8	39.1	41.1
Fiscal Revenues	35.3	45.5	44.3	45.2	46.9
Algeria, Iran, Syria, Yemen Fiscal Expenditure	38.7	59.4	72.5	93.2	92.8
Fiscal Revenues	45.5	60.1	70.5	84.6	93.5
Bahrain, Kuwait, Oman, Saudi Arabia, UAE Fiscal Expenditures	57.9	158.6	188.5	236.0	244.9
Fiscal Revenues	66.4	151.6	162.0	213.3	213.8

Source: World Bank, *MENA Economic Developments and Prospects 2005: Oil Booms and Revenue Management*, pp. 73–74.

Equity markets in the region boomed as more funds were invested locally rather than outside the region. Remittances received did not grow significantly because of the increased employment of non-Arab workers. In the 1990s remittances averaged $11 billion a year; in 2004 they were estimated at $13.2 billion. Tourism within the region did increase, with more being spent by richer Arabs in Egypt, Jordan, and Lebanon.

Tourism and remittance revenues, which averaged nearly $9 billion in the 1990s, rose to an estimated $16.5 billion in 2004.[7]

Table 6.4 provides a summary of the effects of the oil price rise on the fiscal and external balances in the Middle East. The region is divided into sub-sectors: Egypt, Jordan, Morocco, and Tunisia are small or non-oil producers; Algeria, Iran, Syria, and Yemen are oil producers with large populations and labor forces; Bahrain, Kuwait, Oman, Saudi Arabia, and the UAE are large oil producers; data on Libya is not included. The projections for 2005 are likely to prove too low as they were made before the sharp price rises that occurred in the first half of 2005. The table charts a huge growth in export revenues, mainly from the oil sector. Between 1999 and 2005, the exports of Egypt, Jordan, Morocco, and Tunisia doubled, although they export relatively little oil. The net effect of the rise in export revenues, after allowing for the growth in imports and other changes, was a large improvement in the current account of the balance of payments. On the fiscal side, there was also a sharp improvement, due to the direct and indirect effects of higher oil prices within and between the countries of the region. The improvement reinforces the situation of 1999, before the current oil revenue boom, whereby the region, including the main groups of countries, was in fiscal balance, with an overall surplus of $17 billion.

➤ HIGHER OIL REVENUES

The world oil and gas markets have become increasingly regionalized over the years. Far East countries have become more reliant on Middle East oil (and gas), while the US and Europe have remained reliant on oil and thus on its largest supplier, the Middle East. The rise in Chinese oil imports has reinforced the trend of Far East dependence on Middle East oil. In 1990, the US and Europe accounted for 43 percent of Middle East oil exports and the Far East and Pacific region for 45 percent. By 2003, the US and Europe accounted for 36 percent and the Far East and Pacific for 56 percent.[8] China and Japan have maintained their links with Iran while China has invested in Sudan.

Certainly one important manifestation of magnified oil revenues lies in the defense sector. Between 1999 and 2003, defense spending in Middle East countries rose by 39 percent, in 2000 dollars: in 1999 it totaled $50.3 billion and in 2003, $70 billion, also in 2000 prices. The Arab countries increased their defense spending from $32.5 billion to $47.6 billion and Iran from $8.4 billion to $12.6 billion, all in 2000 prices and exchange rates.[9] This was much faster than the world increase of 26 percent, and also much faster than the 15 percent increase recorded by the region in the preceding five years. As a result of these developments, the Middle East's share of world military spending rose from 6.2 percent in 1995 to 7.2 percent in 1999 and to 8 percent in 2004. While enabled by higher oil revenues, regional developments help to explain these trends. First, the Palestinian intifada that began in 2000 affected Israeli defense spending, as did preparations for the Iraq war of 2003. Growing Iranian fears of the US resulted in an increase in its defense allocations. Defense spending rose because of foreign aid (in the cases of Israel, Egypt, and Jordan) and rising oil revenues. In 2004, concern about domestic terrorism led to a large increase in internal security spending in Saudi Arabia.

Table 6.5 shows how the value of arms contracts with major suppliers outside the region and deliveries from them to the Middle East developed between 1999 and 2004. Between the periods 1996–99 and 2000–3 the value of both agreements signed and arms delivered fell significantly, reflecting conditions at the time of signing. The rise in oil revenues in the Middle East OPEC states began in 1999 and became really significant in 2000. Oil revenues then fell in 2001 and 2002 before rising sharply in 2003–5. It is therefore likely that the effects of higher oil prices will only show up in data for 2004 onwards. According to the World Bank, as Middle Eastern countries have been saving a large share of the increased revenues, the volume of new arms orders may not prove to be large.[10] It takes time for new orders to be placed and their volume will be affected by regional developments.

Table 6.5 Middle East Arms Transfer Agreement and Deliveries ($ million)

| | Agreements | | Deliveries | |
	1996–1999	2000–2003	1996–1999	2000–2003
Algeria	1,500	500	1,200	700
Bahrain	600	400	300	600
Egypt	6,800	6,800	4,400	5,400
Iran	1,700	500	2,000	600
Iraq	0	200	0	200
Israel	4,700	5,200	4,500	3,200
Jordan	700	1,000	300	600
Kuwait	900	2,200	4,400	2,100
Lebanon	100	0	200	0
Libya	700	500	200	400
Morocco	600	100	600	300
Oman	300	1,200	600	100
Qatar	800	0	1,800	0
Saudi Arabia	6,000	3,400	37,200	3,900
Syria	500	300	500	200
Tunisia	0	0	200	0
UAE	7,600	8,100	5,100	2,600
Yemen	700	600	400	600
Total Arab	27,260	25,120	57,400	36,700
Total Arab, excluding Egypt and Jordan	19,460	17,320	52,700	30,700
Total ME OPEC	19,200	15,400	51,900	29,500
Total Middle East	**33,660**	**30,820**	**63,900**	**40,500**

Source: Richard F. Grimmett, *Conventional Arms Transfers to Developing Countries, 1996–2003* (Washington DC: Congressional Research Service Report for Congress, 2004).

An interesting trend regards surplus funds. According to the IMF, oil producers have saved most of the export windfall that they received in recent years.[11] Between December 1999 and September 2004, the volume of funds deposited by six Middle East oil producers in banks in the main industrialized countries that are members of the Bank for International Settlements (BIS) increased by 46 percent. The BIS noted that unlike earlier periods when oil revenues rose rapidly, the recent period was not marked by large increases in those deposits (table 6.6).[12]

Table 6.6 BIS Liabilities to Middle East Oil Producers, 1999–2004 ($ billion)

	December 1999	December 2000	December 2001	December 2002	December 2003	September 2004
Iran	8.1	15.9	14.4	16.1	17.7	21.4
Kuwait	15.6	19.3	20.4	22.0	21.6	24.9
Oman	2.9	4.4	5.1	5.3	5.4	5.1
Qatar	3.4	5.2	5.5	5.0	5.5	4.5
Saudi Arabia	49.9	60.3	51.3	51.2	48.2	77.6
UAE	41.2	40.7	56.2	51.5	49.3	43.6
Total	**121.1**	**145.8**	**152.9**	**151.1**	**147.7**	**177.1**

Source: Bank for International Settlements, *Quarterly Review: International Banking and Financial Market Developments*, March 2003 and March 2005, BIS.org.

Saudi data shows that surplus funds were invested in securities (stocks and bonds) rather than in deposits abroad. Between 1999 and 2004, the Saudi Monetary Authority (SAMA) recorded an 84 percent increase in assets, from $55 billion to $100 billion.[13] Kuwait's central bank foreign assets rose from $4.5 billion to $8.7 billion during the same period.[14] Most of Kuwait's surplus was invested in the Fund for Future Generations, run by the Kuwait Investment Authority, which does not disclose information about its finances. Between 1999 and 2004, the foreign reserves of the UAE central bank increased by nearly 50 percent.[15] Qatar's central bank foreign assets and balances with foreign banks rose threefold between 2000 and 2004.[16] Between 1999 and 2004, the Bahrain Monetary Authority's foreign reserves rose by 53 percent.[17] During the same period, Oman's central bank recorded a 27 percent rise.[18]

The fact that more funds were deposited at home rather than abroad was due to political as well as economic factors. Tensions between the two countries have discouraged Saudis from placing funds in the US. The development of the financial system since the first oil boom has also made domestic deposits a far more inviting option. In the unlikely event of sanctions, the fact that fewer Saudi funds are in American banks means that the US ability to pressure Saudi Arabia will be reduced.

➤ HIGHER OIL PRICES AND STATE ECONOMIES

The rise in oil prices and revenues has caused an increase in GDP in oil producing countries. As oil production is owned by the state, the state's revenues have risen and it has been able to spend more on consumption and investment. This in turn has had real effects on the private and non-oil sectors. The balance of payments has been strengthened as have foreign reserves, and foreign and domestic debt levels have been reduced.

The sections below examine the effects of higher oil prices on three groups of countries: the large producers, smaller producers, and importers.

➤ Saudi Arabia

Saudi Arabia, the world's largest oil producer, has benefited most from the rise in oil prices. As shown in table 6.2, its oil export revenues rose from $45 billion in 1999 to $115 billion in 2004. This massive increase has had major effects on the balance of payment, the government's budget, the national debt, GDP, and investment. In response to the rapid rise in prices in the first quarter of 2005, Saudi Arabia announced that it would speed up the second stage of the Harudh field development to increase production capacity by 350,000 b/d by the end of 2005 instead of by the end of 2006.[19]

In 2004, the current account of the balance of payments recorded a surplus of $51.5 billion, a 73 percent increase from 2003 and over four times the surplus recorded in 2002. The government had a budget balance of $26 billion, equal to 10.5 percent of GDP, compared to a surplus of $9.7 billion, or 4.5 percent of GDP in 2003. In 2004, government debt fell by 7 percent to $163 billion, or 66 percent of GDP, compared to 83 percent in 2003 and 119 percent in 1999. GDP increased by 5.3 percent in 2004, compared with 6.4 percent in 2003, but growth in the non-oil private sector accelerated from 3.4 percent in 2003 to 5.7 percent in 2004. This was largely the result of low interest rates, the liberalization of the telecommunications sector, a strong stock market, and high oil revenues. The latter permitted the government to boost spending on the infrastructure, which was suffering from years of underinvestment. In 2004, the GDP was estimated to have increased by 5.3 percent and the surplus on the balance of payments came to $51.5 billion.[20]

Saudi Arabia faces serious long-term socio-economic problems, including high rates of unemployment and one of the world's fastest population growth rates. These factors mean that there is strong pressure for public spending. It also faces serious security threats, terrorism included. In response, the government was reported to have increased security spending by 50 percent in 2004, from $5.5 billion in 2003, which was in addition to the defense budget.

Per capita oil export revenues remained far below high levels reached during the 1970s and early 1980s. In 2004, Saudi Arabia earned around $4,462 per person, versus $22,174 in 1980. This 80 percent decline in real per capita oil export revenues since 1980 was in large part due to the fact that the population has nearly tripled since 1980, while oil export revenues in real terms have fallen by over 40 percent. Meanwhile, Saudi Arabia has experienced nearly two decades of large budget and trade deficits, the effects of the 1990–91 war with Iraq, and a public debt of around $175 billion. On the other hand, Saudi Arabia possesses foreign assets estimated at around $110 billion.[21]

➤ Iran

In the period 2000–5, Iran's economy grew by almost 6 percent a year. This was the result of higher oil prices, higher oil production in 2004–5, economic reforms, and the strength of the non-hydrocarbon sector. The share of oil and gas in total exports fell between 2000–1 and 2004–5 from 70 percent to an estimated 66 percent as a result of growth of other exports. Between 2000–1 and 2004–5, the non-hydrocarbon sector grew by an annual average of 6.1 percent while the hydrocarbon sector grew by only 2.1 percent. The slow growth of the latter stemmed from the modest overall increase in oil output and the fact that gains were mainly due to rising prices.

The financial effects of rising oil revenues were significant. The balance of payments was strengthened by higher exports, although the current account surplus declined between the financial years 2001–2 and 2003–4. In 2000–1, the current account surplus equaled 5.3 percent of GDP. In 2004–5, the estimated surplus was 1.5 percent. Similarly, the budget surplus, excluding hydrocarbons, came to 1.8 percent of GDP. There is evidence that revenues were being used to fund subsidies, public sector employment, and imports of refined petroleum products, which were in short supply in Iran. The Oil Stabilization Fund (OSF) that was set up to cushion the budget and the economy from fluctuations in oil revenues targeted these current needs.[22]

Hydrocarbon product prices in Iran were much lower than their international level. In 2003–4, the implicit subsidy came to $14.2 billion, equal to 10.4 percent of GDP. This was made possible by Iran's oil and gas wealth. This huge subsidy encouraged domestic consumption, then used by government as justification for investing in nuclear power development. The World Bank and the IMF have recommended phasing out domestic energy subsidies for economic reasons, but there has been little movement on this in Tehran. The non-oil sector has recorded fast growth rates in recent years made possible in part by higher oil revenues. These enabled importing more of the inputs that the economy relies on. Oil revenues have helped to fund a large increase in investment that has reduced unemployment and boosted the economic growth rate.[23]

Iran is better placed to withstand international sanctions from a financial point of view: its state budget, balance of payments, foreign debt, and reserves are all much healthier than they were in 1999. It is, however, short of technology for its oil and gas fields, as well as for other sectors. Close relations with Russia and China will not solve this problem.

➤ UAE

Estimates show that the UAE is the third largest OPEC export earner after Saudi Arabia and Iran. Predictably, then, the UAE economy has benefited from high oil prices and revenues and has accumulated large surpluses on the current account of the balance of payments and the government budget. The consolidated government balance, coving the federal government and the emirates' governments, improved from a deficit equal to 10.3 percent of GDP in 1999 to a surplus of 8.1 percent. The current account surplus rose from 0.9 percent of GDP in 1999 to 6.3 percent in 2004.[24] The non-oil and private sectors have developed strongly as a result of the liberal economic policies that have long been in force, especially in Dubai.

➤ Kuwait

Kuwait is one of the most oil-reliant states in the Gulf. Oil revenues account for about 90 percent of government income and 40–50 percent of the country's GDP. Increased oil prices since early 1999 have had positive implications for Kuwait's financial, budgetary, and economic situations. In the fiscal year 2004–5 to March 2005, Kuwait had a budget surplus of some $10 billion, the sixth straight year of budget surpluses. Oil export revenues are estimated to have doubled between 2002 and 2005 (table 6.2). The economy grew by about 5.5 percent in 2003 and 2004 and was expected to grow at a similar rate in 2005.

Kuwait maintains large reserves in the Reserve Fund for Future Generations (RFFG). These were estimated at around $80 billion at the end of 2004. Until December 2004, 10 percent of oil revenues were put into the Fund; since then 15 percent has been allocated.[25] Since the Gulf War of 1991, government investment has been concentrated in the oil sector. Kuwait plans to spend billions of dollars to expand its oil export terminal, refining sector, and production capacity.

➤ Libya

Libya's economy is benefiting from the lifting of UN sanctions in 2003 and of US sanctions in 2004. After years of stagnation, the GDP grew by about 5 percent in 2003 because of the rise in oil revenues.[26] In 2004, oil revenues rose by 33 percent and so GDP is likely to have increased at 8 percent. In 2003, hydrocarbons accounted for nearly 50 percent of the national income, compared with 32 percent in 2000. In 2004 they were estimated to have accounted for just over 52 percent. In the period 2000–3, hydrocarbons accounted for 97 percent of exports. The increase in hydrocarbon income has resulted in a fiscal surplus equal to 10.5 percent of GDP in 2003. In 2003 Libya made a payment of $1.1 billion as part of the Lockerbie settlement and in 2004 was due to make the final payment of $1.6 billion. These payments were made possible by a squeeze on investment allocations in the economy, despite the increase in oil and gas revenues.[27]

In February 2004, following Libya's declaration that it would abandon its weapons of mass destruction programs and comply with the Nuclear Non-Proliferation Treaty, the United States cancelled a ban on travel to Libya. It also authorized US oil companies with pre-sanctions holdings in Libya to negotiate their return to the country if and when the US lifted economic sanctions. In April 2004, the United States eased economic sanctions against Libya, permitting US companies to buy or invest in Libyan oil and products. US banks and other financial service providers were given permission to support these transactions. At the same time, Libya's state-owned National Oil Corporation (NOC) announced that it delivered its first shipment of oil to the United States in over twenty years. In June 2004, the United States and Libya resumed diplomatic relations that were severed in May 1981. In September 2004, President Bush revoked most remaining US sanctions against Libya. The US also ended the ban on imports of oil products refined in Libya, and freed Libyan assets in the US of restrictions.

Libya was the first Middle Eastern state to nationalize its oil industry, and it is therefore significant that it began trying to sell 50–60 percent of shares in Tamoil, the state-owned refining and gas station operator.

➤ Algeria

Algeria's economic growth rate accelerated to 7 percent in 2003 and 5.5 percent in 2004 largely as a result of the oil price boom, but also because of a good harvest in

2003. The balance of payments recorded large surpluses on the current account of $8.8 billion (just over 13 percent of GDP) in 2003 and $12.7 billion (15.5 percent of GDP) in 2004. As a result, significant amounts of foreign debt were repaid and, as a percentage of GDP, it fell from 58 percent in 1999 to a projected 20 percent in 2005. The foreign exchange reserves rose from $11.9 billion in 2000 to an estimated $42.3 billion in 2004.[28]

Algeria has continued to suffer from extremely high unemployment, and with the backing of the IMF, the government has agreed to restructure the economy along more market lines. In the hydrocarbon sector, however, it has backtracked. The hydrocarbons reform bill that was designed to restructure the state-owned Sonatrach and state gas and electricity utility Sonelgaz along more commercial lines has not moved forward. Under the reform plan, Sonatrach would have been partially privatized, with a significant stake remaining in government hands. High oil and gas revenues have reduced the government's need to push forward with such changes.

The fall in unemployment that was recorded in recent years was due to government spending on reconstruction following the 2003 earthquake. It was funded by higher oil revenues but has not created long-term jobs.[29]

➤ Syria

Syria is heavily reliant on exports of oil that were, until the 2003 war in Iraq, boosted by illegal Iraqi exports. These have now ended and Syria's domestic production continues to fall. If reports that production dropped to less than 500,000 b/d were correct, the result is a drop of over 15 percent since 1999. At the same time, high international oil prices have boosted income in recent years and helped to disguise the underlying economic problems facing the country. The economic system is a relic from the socialist phase that dominated many Arab countries from the 1950s, but has since been abandoned by most of them. Syria, however, has not disbanded the mass of political-bureaucratic controls that suffocate private initiative, although private sector entrepreneurs close to the regime operate public-private sector joint ventures. The boundaries between the public and the private sectors are far from clear, an amorphousness that serves the regime well. Syria has a large agricultural sector that also suffers from government controls and inadequate technology. It does, however, provide more of the country's needs than the agricultural sector of other countries.

While the population has been growing by 3 percent, the number of young people entering the labor force, or at least trying to find work, has risen much more quickly. According to official sources, Syria has a 12 percent unemployment rate, with over 550,000 out of work. The reality is much worse, quite possibly double the number admitted by the government. Some 80 percent of the unemployed were up to thirty years of age.[30]

Syria's attempts to counter the decline in oil exports by intensifying oil exploration and production, and move from oil-fired to natural-gas fired electric power, require foreign investment, which has become increasingly difficult to obtain given the country's international isolation.

➣ Egypt

Egypt exports and imports oil and on balance is a small exporter. In the financial year 2003–4 (to 30 June 2004), oil accounted for 37 percent of its merchandise exports. The surplus of oil exports over imports increased by 58 percent between 2002–3 and 2003–5 and is estimated to have risen further since then.[31] The increase in net oil revenues was one of the factors that helped to accelerate GDP growth from 1.8 percent in 2003 to an estimated 2.7 percent in 2004 and 3.8 percent in 2005.[32]

➣ Iraq

The economic recovery in Iraq in 2004 was dominated by oil, which accounted for 75 percent of GDP and 97 percent of exports and government revenues. Oil production in 2004 was 54 percent higher than in 2003, reflecting both a recovery from the war and the continuing effects of terrorism. The northern oilfields have been much more affected by terrorism that those in the south. At the end of 2004 and the beginning of 2005 the activities of the oil ministry, crucial to the rehabilitation of the sector, were slowed by political uncertainties as to who would be appointed as minister. In April 2005, output was 2.1 mb/d, much lower than the US Coalition Provisional Administration's forecast of 3.5 mb/d.[33] Revenues rose sharply as a result of the increase in production and the rise in prices: in 2004 they came to $20 billion compared with $7.5 billion in 2003. This enabled the government to increase payments to civil servants, soldiers, and the police, thereby helping to fund a boom in consumption. As the Iraqi economy moved from the rigid controls of the Ba'athist regime to a US-advocated free trade system, imports increased rapidly.

Partly as a result of the very low base in 2003, the economy grew by about 50 percent in 2004. It is forecast to grow by 17 percent in 2005. Spending on the oil sector was estimated at $3.1 billion in 2004 and forecast at $4.3 billion in 2005. Domestic oil product prices were much lower than world levels, implying a subsidy of some $7 billion or nearly 35 percent of GDP. The government has announced its intention to increase domestic prices over the coming years. Government oil revenues came to 77 percent of GDP in 2004 and were forecast at 75 percent in 2005. The economy has also benefited from large-scale US spending both on the military and on civilian reconstruction. As a result of post-war reconstruction, Iraq's balance of payments experienced large deficits. In 2004 and 2005, the current account deficit was $10 billion.[34]

➣ Israel

Between 1999 and 2004, Israel's fuel bill increased by 114 percent in current dollar terms, while the volume of fuel imports increased by only 7 percent. The fuel bill was dominated by imports of crude oil and coal, but also included some other refined products. In 1980, Israel imported $2.1 billion worth of fuels that accounted for 26 percent of its total imports; in 1990, it imported $1.5 billion, equal to 10 percent of total imports. In 2004 they accounted for 11 percent (table 6.7). Current levels represent a much smaller share of total imports than at the peak in 1980 and are therefore a much lighter burden on the economy.

As a result of the recession that prevailed during most of the period, the current account of the balance of payments strengthened between 1999 and 2004. Between 1999 and 2002 the average annual deficit was $1.6 billion. In 2003 and 2004, there were small surpluses. The increased cost of fuel was thus offset by reductions in other imports. If the economy had grown faster, then fuel imports would have been larger and the effect on the balance of payments greater. On the other hand, a faster growing economy would have created the means to remit those higher costs.[35]

Table 6.7 Israel: Fuel and Total Imports, 1980–2004 ($ billion)

Year	Total Imports	Fuel Imports	Fuel as share of total percent
1980	8.1	2.1	26.0
1990	15.0	1.5	10.0
1998	26.9	1.8	6.7
1999	31.3	2.1	6.9
2003	33.6	3.7	11.0
2004	40.5	4.5	11.1

Source: <www.cbs.gov.il/fr_trade/>.

➤ Jordan

The loss of Iraqi oil that was supplied at subsidized prices by Saddam Hussein affected the Jordanian economy in 2003 as did the loss of Iraqi markets for Jordanian manufacturers and traders. The economy has, however, withstood these blows as well as the rise in international oil prices quite well. It has been helped by a process of reorientation towards the West made possible by the operation of Qualified Industrial Zones in cooperation with Israel and a free trade agreement with the US.

Jordan has benefited from some of the indirect effects of the oil revenue boom in the Arab world as well as changes in conditions in Iraq. Some companies relocated from Iraq to Jordan and others were established in Jordan to supply the Iraqi market. As a result exports to Iraq increased, as did the level of economic activity in Jordan. Manufactured exports to Iraq began to recover in 2004, following their decline in 2003 as a result of the war. Tourism revenues have risen, as have remittances from Jordanians working in the Gulf, which have helped to offset the increase cost of oil imports. The economy grew by an annual average rate of 4.3 percent in the years 2000–4 and provisional figures for 2004 show a further increase.[36]

➤ Turkey

Turkey's fuel import bill increased sharply as international oil prices rose. In 1999, it imported $5.3 billion of fuel accounting for 13 percent of total imports. In 2000, this had risen to $9.5 billion or 16.8 percent, despite the fact that the economy had gone into deep recession. The recession reduced demand for all imports but in 2001, fuel accounted for a new height of 20 percent of total imports. By 2003, with economic growth of just under 6 percent, fuel imports came to $11.4 billion, 16.6 percent of total imports.

Between 1999 and 2003, Turkey's annual fuel bill more than doubled in current dollars, but since 2001 the share in total imports has declined.[37]

➤ THE STRATEGIC DIMENSIONS OF HIGH OIL PRICES

High oil prices have transformed government budgets and balance of payments of the oil producers in the Middle East. They have also raised economic growth rates, though not by large amounts. The imperative of economic reform and the diversification of the Middle East economy away from hydrocarbons remains, but the incentives have been reduced by the upsurge in oil revenues. Middle East oil producing countries need to increase employment but this cannot be done in the oil and gas sectors. They therefore must build or expand other employment-intensive sectors, the most important of which is the manufacturing industry. The rise in oil wealth discourages this for several reasons. First, governments have more funds that they can distribute in different forms of welfare payments to the unemployed and or the poor. Second, they can more easily afford to pay the bloated public sector wages that are a function of large civilian and military employment in the public sector.

Although the inflow of oil revenues would strengthen the exchange rate in oil producing countries, their links to a falling dollar have meant that they have not, in recent years, suffered from the so-called "Dutch Disease" effect (lower non-hydrocarbon exports and higher imports). Insofar as the effects of higher oil prices on the world economy have been smaller in the last few years than the oil price jumps of the 1970s and 1980s, these effects are more likely to persist.[38]

Middle East oil producers have not regained the strategic and economic position that they had in the mid-1970s and early 1980s *vis-à-vis* the West. At that time, the sudden accumulation of massive increases of oil wealth was accompanied by huge investments in economic and social infrastructure and as a result the Middle East accounted for 4.4 percent of world imports in 1980. In 2003, it accounted for only 2.5 percent. In 2005, the estimated oil revenues of Middle East OPEC states came to $280 billion, in 2004 prices. In 1980, on the same basis, they peaked at $443 billion. They have thus reached 63 percent of their peak, after falling to a low of $95 billion in 1998, or 21 percent of the peak. The population increased, and as a result, the 2005 per capita level was only about 32 percent of its peak 1980 level. Purchasing power, which translates into imports, has therefore become much weaker than it was twenty-five years ago and so the region's importance on international markets has declined.

There are several strategic implications regarding the current level of oil production, prices, and revenues. The world economy remains very reliant on a resource that is largely located in the Middle East. As a result, what happens in the region is of much greater importance than what happens in other parts of the developing world that lack oil and gas. Although many of the oil producers lack the political agenda that they had in the 1970s (e.g., using the oil weapons against those who supported Israel during the Yom Kippur War), OPEC has survived as a cartel that has restricted production and helped to push up prices. Since 1999 this was possible because of strong demand. Most forecasts suggest that the emphasis on Middle East oil supplies will increase because that is where most of the world's oil reserves are located.

At the same time, the generation of petro-dollars and petro-euros is unlikely to have the same kind of impact on the international economy that they had in the period 1973–85. Although considerable surpluses have developed in the balance of payments and budgets of many Middle East oil producers, their financial needs have grown as a result of population growth. Furthermore their surpluses are much smaller relative to the total volume of funds in the international system than they were in 1973–85.

Rising oil revenues reinforce the position of the state in the economy and society: it is governments that receive oil revenues. They then redistribute them to different sectors without the accountability that usually accompanies reliance on taxation. This has adverse consequences for political and economic liberalization because it reduces the need for reform. It also makes large amounts of money available for spending on unproductive civilian investment projects and on arms.[39] Since 1999, however, the oil producers have earned enough to fund investment in refining, pipelines, shipping, and exploration. This is in sharp contrast to the situation that prevailed in much of the 1990s, when oil prices were low and the oil producing states lacked the financial means (as well as the technical expertise) to carry out investments on the required scale.[40] At their spring 2005 joint meeting the IMF and World Bank expressed concern about the lack of spare capacity in oil production, as did the International Energy Agency.[41] Whether they perceive it as in their interest to increase oil production capacity remains to be seen.

As a result of several years of high oil prices and no prospect of change in the short term, the international economy is likely to move further towards greater exploitation of alternative energy sources, especially outside OPEC, be they Canadian tar-sands, bio-fuels, or nuclear power. The high cost of energy will encourage China, India, and other developing countries to reduce the energy intensity of their production. This will take years to be felt, but when it happens it could have a significant effect on the price of oil, as it did in the mid-1980s.

On the security dimension, the oil exporting countries are accumulating reserves that could be used to increase arms imports or to fund purchases by other countries in the region such as Syria. Although there is no evidence yet of such a trend, a change in geo-political conditions in the region could yield these results and enable these countries to have "guns and butter." And finally, the foreign policies of major oil consuming states *vis-à-vis* the Arab–Israeli conflict have not been affected by the rise in oil prices as they were in the 1970s. This is because of the widespread realization that the producers need the consumers as much as consumers need producers. Hence the oil price rise has been seen as an economic issue rather than as a political one. The EU, for example, which has adopted a critical stance towards Israel for many years, has not changed its policies in the period of rising oil prices. China, India, and other developing countries that have close connections with Iran and Arab oil producers have not been willing to sacrifice their relations with Israel in order to obtain oil. Insofar as this has been true in a period of tight oil markets, it is unlikely to change if and when conditions in oil and energy markets ease.

Notes

1   Paul Rivlin, *World Oil and Energy Trends: Strategic Implications for the Middle East*, JCSS Memorandum no. 57 (Tel Aviv: Jaffee Center for Strategic Studies at Tel Aviv University, 2000), p. 19.

2   IMF, *World Economic Outlook*, April 2005, p. 157.

3   IEA, *Analysis of the Impact of Higher Oil Prices on the World Economy*, March 2004, <www.iea.org>, p. 5.

4   IMF, *Global Financial Stability Report: Market Developments and Issues*, April 2005, p. 41.

5   IEA, *Oil Market Report*, February 2005, pp. 14–15.

6   EIA, and UN 2004 World Population Prospects, medium variant.

7   World Bank, *MENA Economic Development and Prospects 2005: Oil Booms and Revenue Management*, <www.worldbank.org>.

8   Calculated from BP, *Statistical Review of World Energy*, 2004, <www.bp.com>.

9   *Stockholm Peace Research Institute Yearbook 2004*, table 10A.3.

10 This conclusion is also reached by the World Bank in *MENA Economic Development and Prospects 2005*, p. 19.

11 IMF, *Oil Market Developments and Issues*, March 2005, p. 27.

12 *Bank for International Settlements (BIS) Quarterly Review*, December 2004, p. 25.

13 Saudi Arabian Monetary Agency (SAMA), *Quarterly Statistical Bulletin*, April 2005, SAMA.gov.sa.

14 Central Bank of Kuwait, *Quarterly Statistical Bulletin*, December 2004, cbk.gov.kw.

15 IMF, 2004 Article IV *Consultation Staff Report, UAE*, no. 04/175 June 2004, <www.imf.org>, p. 24.

16 Central Bank of Qatar, *Quarterly Statistical Bulletin*, December 2004, vol. 23, no. 4.

17 Bahrain Monetary Authority, *Quarterly Statistical Bulletin*, December 2004.

18 Central Bank of Oman, *Quarterly Bulletin of Statistics*, December 2004.

19 *Financial Times*, April 16–17, 2005.

20 Saudi American Bank, *Saudi Arabia's 2005 Budget, 2004 Performance*, <www.samba.com/investment/economywatch/pdf/2005Budget.pdf>, p. 1.

21 EIA, *Country Analysis Brief, Saudi Arabia*, January 2005, <www.eia.doe.gov/emeu/cabs/saudi.html>.

22 *Middle East Economic Digest* (MEED), January 7–13, 2005.

23 *MENA Economic Development and Prospects 2005*, pp. 8–10.

24 IMF, Article IV *Consultation and Staff Report, UAE*, no. 04/174, 2004.

25 IMF, Article IV *Consultation Staff Report, Kuwait*, no. 04/186, June 2004.

26 EIU, *Country Report: Libya*, January 2005, p. 12.

27 IMF, Article IV Consultation Staff Report, Libya, no. 05/83 2005, p. 26.

28 IMF, 2004 Article IV *Consultation, Staff Report Algeria*, 2005, pp.7–11.

29 *MENA Economic Development and Prospects 2005*, p. 6.

30 EIU, *Country Report, Syria*, February 2005, pp.12–13, 21–26.

31 EIU, *Country Report, Egypt*, February 2005, p. 37, and *Country Report Egypt*, November 2004, p. 37. Note that the data is incomplete.

32 EIU, *Country Report, Egypt*, February 2005, p. 13.

33 *Financial Times*, 15 April 2005.

34 IMF, *Country Report, Iraq*, 04/325, September 2004 and *MENA Economic Development and Prospects 2005*, p. 16.

35 Bank of Israel, *Annual Report 2004*, and Israel, Central Bureau of Statistics.

36 World Bank, "The Hashemite Kingdom of Jordan," *Quarterly Report of the Jordan Country Unit*, 3rd Quarter, 2004, and *MENA Economic Development and Prospects 2005*, p. 20.

37 IMF, Staff Report no. 02/264, December 2002; 03/324 October 2003 and no. 04/227, July 2004.

38 *MENA Economic Development and Prospects 2005*.

39 Paul Rivlin, "Arab Economies and Political Stability," in Paul Rivlin and Shmuel Even, *Political Stability in Arab States: Economic Causes and Consequences*, JCSS

Memorandum no. 74 (Tel Aviv: Jaffee Center for Strategic Studies at Tel Aviv University, 2004), pp. 9–26.

40 <http://www.iea.org/textbase/npsum/weiosum.pdf>.

41 *Financial Times*, April 16–17, 2005 and April 5, 2005.

Chronology of Major Events, July 2004 – June 2005

9 July 2004	The International Court of Justice declared the construction of Israel's separation fence in the West Bank to be in violation of international law.
21 July 2004	The UN General Assembly passed a resolution demanding Israel's acquiescence to the International Court of Justice ruling that the separation fence in the West Bank be dismantled.
26 July 2004	More than 100,000 people joined hands in a human chain in Israel, in protest of Israel's planned withdrawal from the Gaza Strip.
31 July 2004	Iranian officials stated that Iran had resumed construction of nuclear centrifuges for uranium enrichment, reversing an October 2003 pledge to Britain, France, and Germany to suspend all uranium enrichment-related activities.
26 Aug. 2004	A peace deal to end clashes in Najaf was reached between coalition forces and the Shiite cleric Muqtada al-Sadr.
29 Aug. 2004	Under Syrian pressure, the Lebanese parliament passed a resolution enabling President Emile Lahoud to remain in office for another three years.
1 Sept. 2004	Iranian officials informed the IAEA of their plans to convert "yellow-cake" uranium into uranium hexafluoride.
2 Sept. 2004	The UN Security Council adopted Resolution 1559 demanding that foreign troops withdraw from Lebanon and that Lebanese sovereignty be respected.
19 Sept. 2004	Iran rejected a demand from the IAEA to freeze its uranium enrichment program.
20 Sept. 2004	US president George W. Bush revoked the trade embargo on Libya.
22 Sept. 2004	Syrian officials announced that Syria began to withdraw its troops from Lebanon.
26 Sept. 2004	Hamas leader Izz al-Deen al-Sheikh Khalil was assassinated in Damascus. Israel subsequently acknowledged its involvement.
30 Sept. 2004	Twenty-three Palestinians and three Israelis were killed during an Israeli raid on the Jabalya refugee camp in the northern part of the Gaza Strip, designed to curb Kassam rocket attacks. This was the first time in the four years of the violence the IDF entered the camp.
2 Oct. 2004	Shiite cleric Muqtada al-Sadr told Iraqi leaders that he was planning to disband his militia and enter the political process.

6 Oct. 2004	The EU executive body ruled that Turkey had made enough progress to merit accession talks.
7 Oct. 2004	The CIA's Iraq Survey Group released a report concluding that Iraq did not have weapons of mass destruction prior to the war.
8 Oct. 2004	At least 29 people were killed in three terror attacks on tourist resorts in Sinai. The largest one destroyed the Taba Hilton, close to the Israeli border. Most of the victims were Israelis; Egyptians and other nationals were also among the victims.
11 Oct. 2004	The EU foreign ministers agreed to lift the arms embargo on Libya.
19 Oct. 2004	Pursuant to UN Security Council resolution 1559, the Security Council called on Syria to withdraw its troops from Lebanon.
20 Oct. 2004	Iran's defense minister, Ali Shamkhani, announced that Iran test-fired an improved version of the Shehab-3 ballistic missile.
20 Oct. 2004	Lebanese prime minister Rafiq al-Hariri and his cabinet resigned. President Emile Lahoud named Omar Karami as the new prime minister.
26 Oct. 2004	After two days of deliberations, the Knesset approved Prime Minister Ariel Sharon's disengagement plan by a solid majority.
2 Nov. 2004	President of the UAE and ruler of Abu Dhabi Shaykh Zayid bin Sultan al-Nahayyan died. His eldest son, Shaykh Khalifa, succeeded him as ruler of Abu Dhabi and was then elected as president of the UAE by rulers of the seven emirates.
11 Nov. 2004	Palestinian Authority chairman Yasir Arafat died in a French military hospital after suffering multiple-organ failure. Mahmoud Abbas was appointed the head of the Palestinian Liberation Organization.
14 Nov. 2004	Iran agreed to suspend its nuclear programs immediately in exchange for European guarantees that it would not face UN Security Council sanctions. The deal was struck with ambassadors of Britain, France, and Germany.
22 Nov. 2004	The Shiite cleric Muqtada al-Sadr issued a *fatwa* forbidding his followers to participate in Iraq's forthcoming elections.
28 Nov. 2004	Jordan's King Abdullah stripped his half-brother Hamza of the title of crown prince in a nationally televised appearance.
30 Nov. 2004	Iran's chief negotiator on the nuclear issue, Hassan Rowhani, said Iran agreed to suspend its uranium enrichment activities for a few months only while negotiations with the Europeans for a long-term accord continued, but would never fully abandon the program.
1 Dec. 2004	Hundreds of thousands of Lebanese took part in a pro-Syrian demonstration in Beirut.
6 Dec. 2004	A group of attackers with explosives stormed the American consulate building in Jidda, Saudi Arabia.
8 Dec. 2004	US military intelligence officials concluded that Iraqi insurgency was directed from Syria to a greater extent than previously thought.
14 Dec. 2004	Mahmoud Abbas, chairman of the Palestinian Authority, stated in an interview to the newspaper *Al-Sharq al-Awsat* that the second intifada was a mistake and that all violence should end.

14 Dec. 2004	Egypt and Israel signed their first strategic partnership accord in trade and industry.
18 Dec. 2004	Under UN pressure to withdraw from Lebanon, Syria redeployed security forces from strategically important positions in and around Beirut.
22 Dec. 2004	Saudi Arabia recalled its ambassador from Libya and asked the Libyan ambassador to leave the country in the wake of reports linking Libyan leader Muammar Qaddafi to an assassination plot on Saudi crown prince Abdullah bin Abd al-Aziz.
30 Dec. 2004	Egypt succeeded in convincing both Saudi Arabia and Libya to stop trading media attacks in an effort to limit tension between the two states.
9 Jan. 2005	Mahmoud Abbas was elected by a wide margin as the Palestinian Authority president, in elections held two months after the death of Yasir Arafat.
9 Jan. 2005	The Sudanese government and rebel leaders signed a peace accord ending Sudan's long civil war.
30 Jan. 2005	Elections to the Iraqi parliament were held, the first since the ousting of Saddam Hussein's regime.
14 Feb. 2005	A car bomb killed former Lebanese prime minister Rafiq al-Hariri. The US blamed Syria for the assassination.
14 Feb. 2005	The Israeli Knesset approved the Compensation Act, clearing the way for Prime Minister Sharon's disengagement plan.
16 Feb. 2005	The Knesset passed the disengagement plan law in second and third votes.
21 Feb. 2005	Israel released 500 Palestinian prisoners, as a gesture towards the Palestinian Authority and its newly-elected president, Mahmoud Abbas. The Palestinian Authority demanded the release of 7000 more prisoners.
24 Feb. 2005	The Palestinian parliament ratified a new government headed by Ahmad Qurei.
28 Feb. 2005	Lebanese prime minister Omar Karami resigned after losing a no-confidence vote in parliament. The opposition demanded an investigation into Hariri's assassination.
26 April 2005	Syria completed withdrawal of its forces from Lebanon, ending a 29-year occupation.
28 April 2005	The Iraqi parliament approved the election of Jalal Talabani as president and Ibrahim Jaafri as prime minister, ending three months of political deadlock.
28 April 2005	Vladimir Putin, president of Russia, visited Israel, the first visit of a Russian leader to Israel.
1 May 2005	Recep Tayyip Erdogan, president of Turkey, visited Israel, the first visit of a Turkish leader to Israel.
2 June 2005	Israel released 398 Palestinian prisoners, the second group released since February.
3 June 2005	Syria test-fired three Scud missiles, two of them Scud D, a version developed in Syria. One of the missiles broke apart in mid-air and its debris fell on Turkish soil.

99

4 June 2005	Mahmoud Abbas, president of the Palestinian Authority, announced the postponement of parliamentary elections, originally scheduled for July 2005.
13 June 2005	Egyptian diplomat Mohamed ElBaradei was reelected for a third term as head of the IAEA, the United Nations' nuclear supervisory organization.
20 June 2005	The party of Saad al-Hariri, son of assassinated Rafiq al-Hariri, won the final round of parliamentary elections in Lebanon.
24 June 2005	Conservative Mahmoud Ahmadinejad was elected as president of Iran in a two-round popular election.
3 0 June 2005	Fouad Siniora was named prime minister of Lebanon.

PART ▶▶▶

II

Military Forces

Introductory Note

For each of the countries reviewed below, only total numbers of main weapon systems are given, without breakdown into further detail. General data on each country is presented, along with data on weapons of mass destruction and on space assets. The data on arms procurement as well as arms sales and military industry appears in a concise format. The material presented below is updated to June 2005.

The table representing the order-of-battle of each country often gives two numbers for each weapon category. The first number refers to quantities in active service, whereas the second number (in parentheses) refers to the total inventory.

Charts representing distribution of weapon systems in three distinct regions of the Middle East follow the reviews of the individual countries. The regions are:

1   Eastern Mediterranean (includes Egypt, Israel, Jordan, Lebanon, Syria, and Turkey)
2   The Persian Gulf (includes Bahrain, Iran, Iraq, Kuwait, Oman, Qatar, Saudi Arabia, and UAE)
3   North Africa and other countries (includes Algeria, Egypt, Libya, Morocco, and Tunisia. To these is added Sudan.)

Detailed data on the region's military forces is available online at the Jaffe Center website: <http://www.tau.ac.il/jcss/balance/index.html>.

Economic Data

The tables on economic data include data on GDP (in current US dollars) and defense expenditure only. Sources for the economic data are EIU Country Profiles, EIU Quarterly Reports, IMF International Financial Statistical Yearbook, and SIPRI Yearbook.

Data on military/defense expenditures in the Middle East is notoriously elusive. Hence it should be regarded primarily as an indication of procurement trends.

Arms Trade and Foreign Military Cooperation

Data on military acquisitions and sales as well as on security assistance and foreign military cooperation is limited to information pertaining to the past five years. The year in parentheses indicates the most recent information about the data.

Note on Symbols

NA	Data not available. This symbol is used in the economic data tables only.
~	The tilda is used in front of a number to denote an inexact number.
+	The weapon system is known to be in use, but the quantity is not known.

YIFTAH S. SHAPIR
JUNE 2005

Review of Armed Forces

1 ALGERIA

Major Changes

- Major General Salih Ahmed Jaid is the new chief of the general staff.
- No other major changes were recorded in the Algerian armed forces.

General Data

Official Name of the State: Democratic and Popular Republic of Algeria
Head of State: President of the High State Council Abd al-Aziz Buteflika
Prime Minister: Ahmed Ouyahia
Minister of Defense: Nureddin Zarhouni
Chief of General Staff: Major General Salih Ahmed Jaid
Commander of the Ground Forces: Major General Ahsan Tafer
Commander of the Air Force: Brigadier General Muhammad Ibn Suleiman
Commander of Air Defense Force: Brigadier General Achour Laoudi
Commander of the Navy: Admiral Taher Yali

Area: 2,460,500 sq. km.
Population: 32,800,000

Economic Data (in US $billion)

	2000	2001	2002	2003	2004
GDP (current prices)	54.5	54.9	55.9	66.2	76.8
Defense expenditure	1.88	1.93	2.1	2.2	NA

Major Arms Suppliers

Major arms suppliers are Russia, which supplied combat aircraft, utility helicopters, and naval missiles. Russia also upgraded major weapon systems supplied in the past by the Soviet Union. The Czech Republic supplied tanks and training aircraft.

The US is becoming a major arms supplier to Algeria. It supplied Algeria with C^3I systems, electronic reconnaissance aircraft, ground radars, and financial aid.

Other suppliers are Belarus, which sold Soviet-made combat aircraft; France, which sold utility helicopters; Ukraine, which sold MBTs; and South Africa, which upgraded attack helicopters.

Foreign Military Cooperation

Type	Details
Forces deployed abroad	Congo (MONUC), Ethiopia, and Eritrea (UNMEE) (2001)
Foreign forces	US (2004)
Joint maneuvers	Italy (2003), US maritime SAR and ASW exercises
Security agreements	France (2000), Iran (2002), Libya (2001), Russia (2001), South Africa (2000), Turkey (2003)

Defense Production

Patrol boats, trucks, and small arms.

Strategic Assets

NBC Capabilities

Nuclear capability
One 15 Mw nuclear reactor, probably upgraded to 40 Mw (from PRC) allegedly serves a clandestine nuclear weapons program; one 1 Mw nuclear research reactor (from Argentina); basic R&D. Signatory to the NPT. Safeguards agreement with the IAEA in force. Signed and ratified the African Nuclear Weapon-Free Zone Treaty (Pelindaba Treaty).

Chemical weapons and protective equipment
No data on CW activities available.
Signed and ratified the CWC.

Biological weapons
No data on BW activities available.
Not a party to the BWC.

Space Assets

Model	Type	Notes
Satellites		
ALSAT-1	Research satellite	Earth monitoring for natural disasters
Future launch		
ALSAT-2	Remote sensing	

Armed Forces

Order-of-Battle

Year	2001	2002	2003	2004	2005
General data					
Personnel (regular)	127,000	127,000	127,000	127,000	127,000
SSM launchers					
Ground forces					
Divisions	5	5	5	5	5
Total number of brigades	26	26	26	26	26
Tanks	900 (1,100)	900 (1,100)	900 (1,100)	900 (1,100)	900 (1,100)
APCs/AFVs	2,110*	2,110	1,980 (2,080)	1,980 (2,080)	1,915* (2,015)
Artillery (including MRLs)	900 (985)	900 (985)	900 (985)	900 (985)	920 (1,000)
Air force					
Combat aircraft	184 (214)	228 (258)	228 (258)	228 (258)	213* (243)
Transport aircraft	41 (46)	41 (46)	40 (45)	40 (45)	41 (46)*
Helicopters	133* (142)	131 (140)	174 (183)	174 (183)	174 (183)
Air defense forces					
Heavy SAM batteries	11	11	11	11	11
Medium SAM batteries	18	18	18	18	18
Light SAM launchers	78	78	78	78	78
Navy					
Combat vessels	26	26	26	26	26
Patrol crafts	16	16	16	16	16

Note: The change in the number of APCs/AFVs, specifically BTR 60 / 80 and OT-64, is due to a change in estimate and not a real change. The numbers of Su-22 combat aircraft, An-12 cargo aircraft, and some types of helicopters have changed based on an estimate that these aircraft were phased out.
* Due to change in estimate

Personnel

	Regular	Reserves	Total
Ground forces	107,000	150,000	257,000
Air force	14,000		14,000
Navy	6,000		6,000
Total	**127,000**	**150,000**	**277,000**
Paramilitary			
National Security Force	16,000		16,000
Republican Guards Brigade	1,200		1,200
Gendarmerie	24,000		24,000

2 BAHRAIN

Major Changes

*   No major changes were recorded in the Bahraini armed forces.

General Data

Official Name of the State: State of Bahrain
Head of State: Amir Shaykh Hamad bin Isa al-Khalifa
Prime Minister: Khalifa ibn Salman al-Khalifa
Minister of Defense: Lieutenant General Khalifa ibn Ahmed al-Khalifa
Commander in Chief of the Armed Forces: Salman bin Hamad al-Khalifa
Chief of Staff of the Bahraini Defense Forces: Major General Rashid bin Abdallah al-Khalifa
Commander of the Air Force: Hamad ibn Abdallah al-Khalifa
Commander of the Navy: Lieutenant Commander Yusuf al-Maluallah

Area: 620 sq. km.
Population: 700,000

Economic Data (in US $billion)

	1999	2000	2001	2002	2003
GDP (current prices)	6.8	8.5	7.9	7.7	8.6
Defense expenditure	0.327	0.321	0.329	0.335	0.456

Major Arms Suppliers

Major arms suppliers are the US, which supplied combat aircraft, helicopters, and tactical ballistic missiles, and the UK, which supplied transport and training aircraft. Bahrain also received air defense systems from Sweden.

Foreign Military Cooperation

Type	Details
Forces deployed abroad	Saudi Arabia (part of GCC Desert Shield rapid deployment force)
Foreign forces	About 1,700 US forces (2004)
Joint maneuvers	Egypt (2001), GCC countries (2001), Jordan (2001), US (2003)
Security agreements	US, Britain, GCC countries

Strategic Assets

NBC Capabilities

Nuclear capability
No known nuclear activity.
Signatory to the NPT.

Chemical weapons and protective equipment
No known CW activities.
Party to the CWC.

Biological weapons
No known BW activities.
Party to the BWC.

Future procurement
GID-3 CW detection system (2002)

Ballistic Missiles

Model	Launchers	Missiles	Since	Notes
ATACMS		30	2002	

Armed Forces

Order-of-Battle

Year	2001	2002	2003	2004	2005
General data					
Personnel (regular)	7,400	8,200	8,200	8,200	8,200
SSM launchers	9	9	9	9	9
Ground forces					
Total number of brigades	3	3	3	3	3
Number of battalions	7	7	7	7	7
Tanks	180	180	180	180	180
APCs/AFVs	277 (297)	277 (297)	277 (297)	277 (297)	277 (297)
Artillery (including MRLs)	48 (50)	48 (50)	48 (50)	48 (50)	48 (50)
Air force					
Combat aircraft	34	34	34	33 (34)	33 (34)
Transport aircraft	2	2	3	3	3
Helicopters	39 (41)	40 (42)	40 (42)	40	48*
Air defense forces					
Heavy SAM batteries	1	1	1	1	1
Medium SAM batteries	2	2	2	2	2
Light SAM launchers	40	40	40	40	40
Navy					
Combat vessels	11	11	11	11	11
Patrol boats	21	21	21	22	22

* Due to change in estimate

Personnel

	Regular	Reserves	Total
Ground forces	6,000		6,000
Air force	1,500		1,500
Navy	700		700
Total	**8,200**		**8,200**
Paramilitary			
Coast Guard and National Guard	2,000		2,000

3   EGYPT

Major Changes

- The Egyptian industry is still occupied with the assembly of more M1A1 Abrams MBTs. This will increase the total number of M1A1 MBTs in Egypt to 880 tanks by 2008.
- The Egyptian military industry will produce an unspecified quantity of the 155mm GH 52 Patria towed artillery (with Auxiliary power unit) under technology transfer agreement with Finland.
- The 26 MLRS firing units ordered from the US are assessed to be already in service.
- The air force is absorbing its 24 new F-16D in the framework of the Peace Vector VI deal. The air force also decided to upgrade its 35 Apache AH-64A to the AH-64D standard, although Egypt will not receive the Longbow radar.
- The air force received two of its five Hawkeye AEW aircraft that are being upgraded to the Hawkeye-2000 standard.
- The air force ordered five An-74 transport aircraft from the Ukraine. These are to be supplied beginning in 2005.
- The navy ordered three King Cobra (formerly Ambassador Mk III) missile patrol boats. The navy is also absorbing six used Tiger missile patrol boats from Germany.
- The navy ordered six Swift Protector fast patrol boats from the US.
- Egypt s plans to acquire the Moray submarines are probably frozen. Meanwhile Egypt is negotiating a possible acquisition of some 206A submarines from Germany.

General Data

Official Name of the State: The Arab Republic of Egypt
Head of State: President Muhammad Husni Mubarak
Prime Minister: Ahmed Nadhif
Minister of Defense and Military Production: Field Marshal Muhammad Hussayn Tantawi
Chief of General Staff: Lieutenant General Hamdi Wahaba
Commander of the Air Force: Maj. Gen. Magdi Galal Sha rawi
Commander of the Navy: Vice Admiral Ahmed Saber Salim

Area: 1,000,258 sq. km. (dispute with Sudan over Halaib triangle area)
Population: 73,300,000

Economic Data (in US $billion)

	2000	2001	2002	2003	2004
GDP (current prices)	97.9	90.4	84.1	71.1	71.9
Defense expenditure*	2.46	2.61	2.75	NA	NA

* Published defense expenditure data apparently does not include annual $1.3 bn foreign military assistance from the US.

Major Arms Suppliers

The major arms supplier is the US (MBTs, MLRS, combat aircraft, attack helicopters, radars, combat vessels, advanced air force and naval armament). Other suppliers include Germany (missile patrol boats), North Korea (ballistic missiles), PRC (training aircraft), Netherlands (AIFVs), Ukraine (upgrading SAMs and tanks, transport aircraft), Belarus (upgrading SAMs and APCs), Russia (upgrading SAMs).

Major Arms Transfers

No major arms transfers were concluded over the last five years.

Foreign Military Cooperation

Type	Details
Foreign forces	US forces as of September 2003 include some 350 soldiers; MFO s soldiers as follows: Australia (25), Canada (29), Colombia (358), Fiji (338), France (15), Hungary (41), Italy (75), New Zealand (26), Norway (4), Uruguay (60), US (687)
Forces deployed abroad	Peninsula shield force (2003), Georgia (UNOMIG), Liberia (UNMIL), Western Sahara (MINURSO), Sierra Leone (UNAMSIL), Democratic Republic of the Congo (MONUC)
Joint maneuvers	France, GCC countries (2001), Germany (2001), Greece (2001), Italy (2004), Jordan (2001), Netherlands, Saudi Arabia (2004), Spain (2001), UK (2001), US (2001)

Defense Production

Ballistic missiles, assembly of American MBTs, artillery pieces. Upgrading of AFVs. Assembly of basic training aircraft. Small patrol boats. Electronics and optronic equipment.

Strategic Assets

NBC Capabilities

Nuclear capability

22 Mw research reactor from Argentina, completed 1997; 2 Mw research reactor from the USSR, in operation since 1961.

Party to the NPT. Safeguards agreement with the IAEA in force. Signed but not ratified the African Nuclear Weapon-Free Zone Treaty (Pelindaba Treaty).

Chemical weapons and protective equipment

Alleged continued research and possible production of chemical warfare agents. Alleged stockpile of chemical agents (mustard and nerve agents).

Personal protective equipment; Soviet type decontamination units; Fuchs (Fox) ABC detection vehicle (12); SPW-40 P2Ch ABC detection vehicle (small quantity).

Refused to sign the CWC.

Biological weapons

Suspected biological warfare program; no details available.

Not a party to the BWC.

Ballistic Missiles

Model	Launchers	Missiles	Since	Notes
SS-1 (Scud B/ Scud C)	24	100	1973	Possibly some upgraded
Future procurement				
Scud C/ Project-T		90		Locally produced
Vector				Unconfirmed
No-Dong		24		Alleged

Space Assets

Model	Type	Notes
Satellites		
NILESAT-1/2	Communication	Civilian
Ground stations		
Aswan	Remote sensing	Receiving and processing satellite images for desert research
Future procurement		
DesertSat	Environmental	Monitoring coastal erosion, desertification, and water resources
EgyptSat 1	Remote sensing	100kg; a sun-synchronous, 668km orbit

Armed Forces

Order-of-Battle

Year	2001	2002	2003	2004	2005
General data					
Personnel (regular)	450,000	450,000	450,000	450,000	450,000
SSM launchers	24	24	24	24	24
Ground forces					
Divisions	12	12	12	12	12
Total number of brigades	49	49	49	49	49
Tanks	~3,000	~3,000	~3,000	~3,000	~3,100
	(3,585)	(3,585)	(3,605)	(3,605)	(3,705)
APCs/AFVs	~3,400	~3,400	~3,400	~3,680	~3,680
	(~5,300)	(~5,300)	(~5,300)	(~4,950)	(~4,950)
Artillery (including MRLs)	~3,530	~3,530	~3,530	~3,530	~3,556
	(~3,570)	(~3,570)	(~3,570)	(~3,570)	(~3,596)
Air force					
Combat aircraft	481 (494)	505 (518)	505 (518)	505 (518)	505 (518)
Transport aircraft	44	48	48	44	47
Helicopters	~225	~225	~225	~230	~230
Air defense forces					
Heavy SAM batteries	109	109	109	109	109
Medium SAM batteries	44	44	44	44	44
Light SAM launchers	105	105	105	105	105
Navy					
Submarines	4	4	4	4	4
Combat vessels	64	62	62	59	59
Patrol crafts	104	104	109	103	103

Personnel

	Regular	Reserves	Total
Ground forces	320,000	150,000	470,000
Air force	30,000	20,000	50,000
Air defense	80,000	70,000	150,000
Navy	20,000	14,000	34,000
Total	**450,000**	**254,000**	**704,000**
Paramilitary			
Coast Guard	2,000		2,000
Frontier Corps	6,000		6,000
Central Security Forces	325,000		325,000
National Guard	60,000		60,000
Border Guard	12,000		12,000

4   IRAN

Major Changes

- Iran is still negotiating the status of its uranium enriching facilities. Iran claims that it is developing an indigenous fuel cycle for civilian use but it is generally perceived to be part of a nuclear weapons program. Although Iran agreed to suspend its uranium enrichment programs it has announced that it intends to reactivate them.
- The Iranian navy received China Cat fast patrol boats from China as well as C-701 ship-borne missiles from China. The Iranian navy also received 25 light attack vessels from North Korea. These include fast light torpedo boats and light semi-submersibles.
- The Iranian navy is also expected to acquire the indigenously built Mowaj corvette and some indigenously built Sina MFPBs.
- It was revealed last year that Ukrainian officials transferred six Kh-55 long-range ALCMs to Iran. These missiles are deemed non-operational, but they may serve as a model for an Iranian missile.

General Data

Official Name of the State: Islamic Republic of Iran
Supreme Religious and Political National Leader (Rahbar): Ayatollah Ali Hoseini Khamenei
Head of State (formally subordinate to National Leader): President Mahmoud Ahmadinejad
Minister of Defense: Mostafa Mohammad Najar
Commander in Chief of the Armed Forces: Major General Mohammad Salimi
Head of the Armed Forces General Command Headquarters: Major General Hasan Firuzabadi
Chief of the Joint Staff of the Armed Forces: Brigadier General Abdol Ali Pourshasb
Commander of the Ground Forces: Brigadier General Nasser Mohammadi-Far
Commander of the Air Force: Brigadier General Karim Qavami
Commander of the Navy: Rear Admiral Abbas Mohtaj
Commander in Chief of the Islamic Revolutionary Guards Corps (IRGC): Major General Yahya Rahim Safavi
Chief of the Joint Staff of the IRGC: Rear Admiral Ali Akbar Ahmadian
Commander of the IRGC Ground Forces: Brigadier General Mohammad Ali Jaafri
Commander of the IRGC Air Wing: Brigadier General Ahmed Kazemi
Commander of the IRGC Naval Wing: Rear Admiral Ali Morteza Saffari

Area: 1,647,240 sq. km. (not including Abu Musa Island and two Tunb islands; control disputed)
Population: 69,100,000 est.

Economic Data (in US $billion)

	1999	2000	2001	2002	2003
GDP (current prices)	55.2	71.9	84.6	117.2	130.2
Defense expenditure	1.64	2.71	3.62	4.34	6.08

Major Arms Suppliers

Major arms suppliers are Russia, which supplied submarines, MBTs, helicopters, combat and transport aircraft, and AD systems. The PRC supplied fast missile patrol boats, transport aircraft, cruise missiles, and AD systems. North Korea supplied and assisted Iran in the production of SSMs. Recently the role of Pakistan as a supplier of nuclear technology surfaced.

Other suppliers include Ukraine, which supplied tanks, transport aircraft, and some ALCMs; Romania, which supplied AD systems; France, which supplied trainer aircraft; and Latvia, which supplied AD systems.

Major Arms Transfers

Iran supplied armament and financial aid to Hizbollah in Lebanon. Arms supplies included MRLs, long range rockets, ATGMs, and shoulder launched SAMs. Some Palestinian organizations received aid that included ATGMs and mortars. Iran cooperated with Syria in the development of ballistic missiles and allegedly in the production of chemical weapons.

Foreign Military Cooperation

Type	Details
Forces deployed abroad	Ethiopia and Eritrea (UNMEE)
Joint maneuvers	India (1998), Italy (2001), Kuwait (proposed naval maneuvers), Oman (observers, 1999), Pakistan (naval maneuvers 2003)
Security agreements	India (2003), Pakistan (2003)

Defense Production

SSMs, tanks, armored combat vehicles, self-propelled guns, towed guns, artillery rockets, anti-tank missiles, attack helicopters, transport aircraft, trainer aircraft, helicopters, patrol crafts, midget submarines, UAVs, AD systems, cruise missiles, guided bombs, radars, fire-control systems.

Note: Some of the weapon systems may be copies of foreign types and not indigenously developed. In addition, some may be only prototypes, which were displayed for propaganda purposes and are not in production.

Strategic Assets

NBC Capabilities

Nuclear capability
One 5 Mw research reactor acquired from the US in the 1960s (in Tehran) and one small 27 kw miniature neutron source reactor (in Isfahan). One 1,000 Mw VVER power reactor under construction, under a contract with Russia, in Bushehr; uranium enrichment facility in Natanz and heavy water production facility in Arak connected to an alleged nuclear weapons program.

Party to the NPT. Safeguards agreement with the IAEA in force. Agreed to sign the IAEA Additional Protocol. Still negotiating the status of its uranium enrichment program.

Chemical weapons and protective equipment
Iran admitted in 1999 that it had possessed chemical weapons in the past.
Personal protective equipment and munitions decontamination units for part of the armed forces.

Party to the CWC, but nevertheless suspected of still producing and stockpiling mustard, sarin, soman, tabun, VX, and other chemical agents. Alleged delivery systems include aerial bombs, artillery shells, and SSM warheads. PRC and Russian firms and individuals allegedly provide assistance in CW technology and precursors.

Biological weapons
Suspected biological warfare program; no details available.

Party to the BWC.

Ballistic Missiles

Model	Launchers	Missiles	Notes
SS-1 (Scud B/ Scud C)	~20	300 Scud B, 100 Scud C	
Shehab-2	+	+	Probably similar to the Syrian Scud-D
Shehab-3	6	~20	
CSS-8	16		
Total	**~40**		
Future procurement			
Shehab-3B			Under development; includes new RV
Fateh-110			Under development

Space Assets

Name	Type	Notes
Ground station		
IRSC	Remote sensing	Multi-spectral remote sensing
Future procurement		
Sina-1	Remote sensing	A 170 kg remote sensing satellite to be launched by a Russian launcher
Safir 313	Research	Micro satellite to be indigenously launched in 2005
Mesbah	Research	To be launched in 2005 in cooperation with Italy
Zohreh	Communication	

Armed Forces

Order-of-Battle

Year	2001	2002	2003	2004	2005
General data					
Personnel (regular)	~520,000	~520,000	~520,000	~520,000	~520,000
SSM launchers	~40	~40	~40	~40	~40
Ground forces					
Divisions	32	32	32	32	32
Total number of brigades	87	87	87	87	87
Tanks	~1,700	~1,700	~1,700	~1,700	~1,620
APCs/AFVs	~1,570	~1,570	~1,570	~1,570	~1,400
Artillery	~2,700	~2,700	~2,700	~2,700	~2,700
(including MRLs)	(~3,000)	(~3,000)	(~3,000)	(~3,000)	(~3,000)
Air force					
Combat aircraft	209 (337)	207 (335)	207 (342)	203 (341)	203 (341)
Transport aircraft	105 (123)	105 (124)	105 (134)	105 (134)	80 (114)*
Helicopters	325 (560)	345 (580)	365 (600)	365 (600)	340 (570)*
Air defense forces					
Heavy SAM batteries	29	29	29	29	29
Medium SAM batteries	+	+	+	+	+
Light SAM launchers	95	95	95	95	95
Navy					
Combat vessels	29	29	29	29	29
Patrol craft	~110	~110	~110	~110	~110
Submarines	3	3	3	3	3

* Due to change in estimate

Personnel

	Regular	Reserves	Total
Ground forces	~350,000	350,000	700,000
Air force	18,000		18,000
Air defense	12,000		12,000
Navy	~18,000		18,000
IRGC ground forces	100,000		100,000
IRGC navy	20,000		20,000
Total	**~520,000**	**350,000**	**870,000**
Paramilitary			
Baseej		2,000,000	

5 IRAQ

Major Changes

- Iraq is in the process of rebuilding its armed forces, aided by coalition forces still present on Iraqi soil.
- These forces include an army and a civilian police force, a National Guard force, a border police, and a facilities protection service. Military hardware is being donated by various countries, including Jordan, Poland, UAE, and US.
- A large occupation force is still present in Iraq. This force is predominantly American and British but personnel from some 20 countries are present in Iraq. Besides military personnel the Americans employ private companies in both logistical and purely military duties.

General Data

Official Name of the State: The Republic of Iraq
Commander of Allied Forces in Iraq: Lt. Gen. George Casey
Head of State: President Jalal Talabani
Prime Minister: Ibrahim Jaafri
Minister of Defense: Saadoun al-Dulaimy
Minister of Interior: Bayan Jabor
Chief of Staff of Ground Forces: Lt. General Amar Bakir al-Hashimi
Commander of the Air Force: Lt. General Kamal al-Barzanji
Commander of Coastal Defense Force: Colonel Hameed Balafam
Commander of National Guard: Major General Muzher al-Mullah al-Rashidi

Area: 432,162 sq. km.
Population: 23,600,000 est.

Economic Data (in US $billion)

	1999	2000	2001	2002	2003
GDP (current prices)	25.3	33.6	29.3	26.8	19.9

Note: Economic data on Iraq is scarce and unreliable.

Major Arms Suppliers

A number of countries donated excess military equipment to Iraq following the establishment of the New Iraqi Army. These included Jordan, which supplied APCs, light reconnaissance and transport aircraft, and utility helicopters; Poland, which supplied utility helicopters; PRC, which supplied patrol boats; UAE, which supplied APCs and light reconnaissance aircraft; and the US, which supplied APCs and small boats.

Defense Production

Since the 2003 war the Iraqi defense industry has not functioned, and it will likely not regain any considerable production capability in the foreseeable future.

After the 1991 Gulf War until the 2003 war, Iraq produced ballistic missiles, UAVs or converting combat aircraft to weapon-carrying UAVs, and upgraded its air defense systems.

Strategic Assets

NBC Capabilities

Nuclear capability
No active nuclear facilities survived.

Party to the NPT.

Chemical weapons and protective equipment
No evidence of remaining chemical warfare capability.

Not a party to the CWC.

Biological weapons
No evidence of remaining biological warfare capability.

Party to the BWC.

Ballistic Missiles

No evidence of remaining ballistic missile capability.

Armed Forces

Order-of-Battle

Year	2001	2002	2003	2004	2005
General data					
Personnel (regular)	432,500	432,500	5,560	9,750	9,750
Ground forces					
Number of battalions			45	60	90
Tanks	2,000	2,000	0	10	20
	(2,400)	(2,400)			
APCs/AFVs	2,000	2,000	0	56	120 (150)
	(2,900)	(2,900)			
Air force					
Reconnaissance aircraft				7	11
Transport aircraft	+	+	0	0	3
Helicopters	360 (460)	370 (460)	0	6	6
Navy					
Patrol crafts	0	1	0	5	5

Note: The number of battalions includes both united belonging to the Ministry of Defense and to the Ministry of the Interior. Not all these unites are fully operational, and most are in various states of training and at different levels of equipment.

Personnel

	Active	Anticipated
Ministry of Defense		
Iraqi National Army	9,750	35,000
Air Corps	150	500
Coastal Defense Force	410	
Iraqi Intervention Force	3,500	6,500
National Guard (formerly ICDC)	37,500	62,000
Ministry of Interior		
Rapid Response Force	1,800	6,000
Police Commando Force	4,000	
Mechanized Police Force	750	1,500
Public Order Police	1,200	
Department of Border Enforcement	18,000	32,000
Paramilitary Forces		
Facility Protection Service	74,000	80,000

Note: The numbers are in constant flux. The data above is as of March 2005.

Foreign Military Personnel

Approximately 138,000 US troops. International Coalition forces includes Albania (200), Australia (850), Armenia (50), Azerbaijan (400), Bulgaria (430), Czech Republic (80), Denmark (530), El Salvador (360), Fiji (155) Georgia (850), Italy (3,100), Japan (550), Jordan (400), Latvia (120), Lithuania (100), Mongolia (180), Poland (1,700), Romania (750), Singapore (33), Slovakia (100), South Korea (3,700), Turkey (4,000), UK (9,000), Ukraine (1,400)

6   ISRAEL

Major Changes

- While the Israeli army is still absorbing its Merkava Mk III MBTs, the new Merkava Mk IV has been presented and has begun entering into service.
- The Israeli army merged two divisional HQs that dealt with security in the West Bank into a single HQ. The change is due to budgetary constraints.
- The Israeli air force chose to implement its option on the F-16I, and ordered 52 additional aircraft. While the first F-16I (Sufa) arrived, aircraft from the second batch will be supplied between 2005 and 2008.
- The first Gulfstream V aircraft ordered by the air force arrived in Israel, where they will be equipped with electronic equipment for its SIGINT missions.
- The Israeli air force deployed the second Arrow BMD battery.
- Due to budget cuts, the air force is considering phasing out older aircraft and closing down some of its squadrons. In addition, the Hawkeye E-2C early warning planes have been deactivated.
- The air force acquired its first advanced JDAM precision guided munitions.
- The air force also received its GROB-120B (Snunit) trainer aircraft. They are operated by a private company.
- Israel launched its new reconnaissance satellite Ofeq 5 to replace the aging Ofeq 3. Besides the military Ofeq series, Israel operates the EROS 1A, which is the first of a projected system of eight civilian reconnaissance satellites. The Eros 1B is scheduled to be launched in 2006.
- The Israeli navy ordered eight new patrol boats six Super Devora Mk II and two Shaldag. The first of these were supplied in 2004.

General Data

Official Name of the State: State of Israel
Head of State: President Moshe Katsav
Prime Minister: Ariel Sharon
Minister of Defense: Shaul Mofaz
Chief of General Staff: Lieutenant General Dan Halutz
Commander of the Air Force: Major General Eliezer Shkedi
Commander of Army HQ: Major General Yiftah Ron Tal
Commander of the Navy: Rear Admiral David Ben Ba ashat

Area: 22,145 sq. km, including East Jerusalem and its vicinity, and the Golan Heights
Population: 6,780,000

Economic Data (in US $billion)

	2000	2001	2002	2003	2004
GDP (current prices)	114.8	113.6	104.2	110.4	114.1
Defense expenditure	8.93	9.03	9.84	9.98	NA

Sources: EIU Quarterly Report, EIU Country Profile, IMF International Financial Statistical Yearbook, SIPRI Yearbook.

Major Arms Suppliers

The US is Israel s major foreign arms supplier, supplying combat aircraft, training aircraft, attack helicopters, helicopters, missile corvettes, tank transporters, SAMs, JDAMs, naval SSMs, MLRS, ATGMs, AMRAAM, SP artillery, and other systems.

Other suppliers include Germany, which supplied Dolphin submarines, training aircraft, NBC detection vehicles, CW protection gear, and Seahake heavy torpedoes. South Africa supplied patrol boats, the Netherlands supplied CW protection gear and assistance in building patrol boats, France supplied training aircraft and CW detectors, and Canada supplied helicopter simulators.

Major Arms Transfers

India is the major Israeli arms receiver. Recently it ordered three Phalcon AEW aircraft in a nearly $1 billion deal. It procured from Israel UAVs, radars, patrol boats, naval SAMs, anti-radar drones, communication equipment, and surveillance systems.

The US procured AGMs, AAMs, digital mapping systems, airborne search and rescue systems, tactical air-launched decoys, flight simulators, mortars, central computers for AFVs, and mine clearing systems. Turkey received AGMs, debriefing systems, aircraft simulators, radars, ECMs, anti-radar missiles, search and rescue systems, and debriefing systems.

Other recipients include Angola (aircraft for ELINT and surveillance, transport helicopters, upgrading of aircraft), Australia (ESMs, APCs radars, night vision equipment, guns for patrol boats), Belgium (UAVs, debriefing systems), Brazil (combat aircraft, avionics suit), Canada (ESMs, OWSs), Chile (AAMs, AAGs, missiles for patrol boats), Cyprus (ECSs), Finland (UAVs, ATGMs, communication equipment), France (debriefing systems), Germany (ATGMs), Greece (EW systems, patrol boats), Italy (laser guided bombs, SAR system, Litening pods, debriefing systems, simulators), Ivory Coast (upgrading of aircraft, UAVs), Mexico (aircraft), Netherlands (ATGMs, C² systems, debriefing systems), Philippines (mini-UAVs), Poland (ATGMs), Portugal (ESMs, debriefing systems), Romania (OWS-25 systems, ground radar systems, upgrading of BMPs, MRLs), Russia (UAVs), Singapore (ATGMs, reconnaissance satellite), South Korea (EW systems, AGMs, anti-radar drones, night vision systems, debriefing systems, radars, satellite reconnaissance equipment, aircraft), Spain (radars), Sri Lanka (attack aircraft, MFPBs, UAVs, radars, ESMs, patrol boats, reconnaissance systems), Taiwan (submarines), Thailand (mini-UAVs, search and rescue systems), UK (UAVs, debriefing systems), Venezuela (radars, Litening pods, SAMs, ESMs).

Foreign Military Cooperation

Type	Details
Foreign forces	Pre-positioning of $200 million worth of stockpiled US military equiment
Cooperation in military training	US and Turkish use of Israeli airfields and airspace for training (2003); Israeli use of Turkish airspace and airfields for training (2003)
Joint maneuvers	Romania (2004), Turkey SAR and joint air force maneuvers (2005), NATO (2005), US (2005)

Defense Production

Major systems produced by Israel include: Merkava MBTs, THEL system, SP AAGs, ATRLs, simulators, Arrow ATBM, ALCMs, AAMs, AGMs, CBUs, TV and guided bombs, radars, UAVs and mini-UAVs, attack UAV, LCTs, MFPBs, PBs, SSMs, ELINT equipment, ESM, EW jammers, command and control systems, night vision devices, satellite launchers, imaging satellites, communication satellites.

Strategic Assets

NBC Capabilities

Nuclear capabilities
Two nuclear research reactors; alleged stockpile of nuclear weapons.*

Not a party to the NPT.

Chemical weapons and protective equipment
Personal protective equipment; unit decontamination equipment.

Fuchs (Fox) NBC detection vehicles (8 vehicles); SPW-40 P2Ch NBC detection vehicles (50 vehicles); AP-2C CW detectors.

Signed but not yet ratified the CWC.

Biological weapons capabilities
Not a party to BWC.

* According to foreign publications

Ballistic Missiles

Model	Launchers	Missiles	Since	Notes
MGM-52C (Lance)	12		1976	
Jericho Mk 1/2/3 SSM*	+			
Total	**+**			

* According to foreign publications

Space Assets

Model	Type	Notes
Satellites		
Amos 1/2	Communication	Civilian
Ofeq series	Reconnaissance	Currently deployed Ofeq 5
Eros	Reconnaissance	Civilian derivative of Ofeq
TechSat	Research	Civilian
Future launches		
MILCOM	Communication	
TECHSAR	Reconnaissance	Equipped intelligence satellite
David	Remote sensing	

Armed Forces

Order-of-Battle

Year	2001	2002	2003	2004	2005
General data					
Personnel (regular)	186,500	186,500	186,500	186,500	186,500
SSM launchers	+	+	+	+	+
Ground forces					
Divisions	16	16	16	16	16
Total number of brigades	76	76	76	76	76
Tanks	3,930	3,930	3,700(3,930)	3,700(3,930)	3,700(3,910)
APCs/AFVs	8,040	8,000	7,710	7,710	6,870
Artillery (including MRLs)	1,348 (1,948)	+	+	+	+
Air force					
Combat aircraft	533 (798)	538 (798)	518 (798)	518 (798)	470 (820)
Transport aircraft	64 (87)	64 (79)	58 (63)	58 (63)	70 (75)
Helicopters	232 (297)	239 (302)	205 (283)	205 (283)	181 (283)
Air defense forces					
Heavy SAM batteries	22	22	22	22	22
Light SAM launchers	~70	~70	~70	~70	~70
Navy					
Submarines	6	5	5	5	3
Combat vessels	20	15	15	15	15
Patrol craft	32	33	33	33	40

Personnel

	Regular	Reserves	Total
Ground forces	141,000	380,000	521,000
Air force	36,000	55,000	91,000
Navy	9,500	10,000	19,500
Total	**186,500**	**445,000**	**631,500**
Paramilitary			
Border Police	7,650		7,650

7 JORDAN

Major Changes

- The American assistance to Jordan was increased to $206 million in 2005.
- The Jordanian army is in the process of absorbing more Challenger I MBTs from Great Britain. Challenger I is replacing the Tariq (Centurion) MBTs. Jordan launched a project to upgrade 180 of its M60 MBTs with new engines and new fire-control systems. Two battalions of these upgraded MBTs have already been equipped.
- Jordan received 100 Ratel 20 IFVs from South Africa and launched a project to upgrade its M113 APCs. The Jordanian military is also introducing into service a variety of locally developed and produced light vehicles.
- The Jordanian air force received eight more F-16 combat aircraft out of an order of 16 aircraft from a USAF drawdown. These aircraft will receive the standard mid-life upgrade (MLU).
- Jordan is transferring various types of its military equipment to the newly formed Iraqi armed forces. These include the BTR-94 and Spartan APCs, C-130 transport aircraft, and UH-1H helicopters.

General Data

Official Name of the State: The Hashemite Kingdom of Jordan
Head of State: King Abdullah bin Hussein al-Hashimi
Prime Minister: Faisal al-Fayez
Minister of Defense: Faisal al-Fayez
Inspector General of the Armed Forces: Major General Abd Khalaf al-Najada
Chief of the Joint Staff of the Armed Forces: Lieutenant General Khalid Sarayrah
Commander of the Air Force: Major General Hussein al-Biss
Commander of the Navy: Commodore Ali Mahmoud al-Khasawna

Area: 90,700 sq. km.
Population: 5,300,000

Economic Data (in US $billion)

	1999	2000	2001	2002	2003
GDP (current prices)	8.1	8.4	8.9	9.5	10.1
Defense expenditure	0.72	0.75	0.76	0.78	0.84

Major Arms Suppliers

Major arms suppliers include the US, which supplied combat aircraft, self-propelled artillery, anti-tank missiles, and radars, and the UK, which supplied MBTs, APCs, heavy transporters, training aircraft, and upgraded light tanks.

Other suppliers include Belgium (APCs), France (helicopters), the Netherlands (self-propelled artillery), South Africa (APCs), Spain (transport aircraft), Turkey (APCs and transport aircraft), Ukraine (APCs), and Canada (upgrading of transport aircraft).

Major Arms Transfers

Jordan sold used combat aircraft to the Philippines; APCs, helicopters and reconnaissance aircraft to Iraq; reconnaissance aircraft to South Africa; and light armored vehicles to Kuwait, Libya, Qatar, and UAE.

Foreign Military Cooperation

Type	Details
Forces deployed abroad	Small contingency force in Afghanistan, Burundi (ONUB), Congo (MONUC), East Timor (UNTAET), Ethiopia and Eritrea (UNMEE), observers in Georgia (UNOMIG), Iraq, Ivory Coast (UNOCI), Kosovo (UNMIK), Liberia (UNMIL), Sierra Leone (UNAMSIL)
Foreign forces	Some US prepositioning of military equipment; UK forces
Cooperation in military training	Turkey (use of facilities and airspace for training of pilots); training for the new Iraqi forces
Joint maneuvers	Egypt, France, Oman, Qatar, Turkey, UAE, UK, US
Security agreements	Saudi Arabia, Turkey, US

Defense Production

Upgrading of tanks, APCs, conversion of IFVs, MRLs, reconnaissance aircraft, and night vision equipment.

Strategic Assets

NBC Capabilities

Nuclear capability
No known capability.

Party to the NPT.

Chemical weapons and protective equipment
Personal protective and decontamination equipment.

No known CW activities.

Party to the CWC.

Biological weapons
No known BW capability.

Party to the BWC.

Armed Forces

Order-of-Battle

Year	2001	2002	2003	2004	2005
General data					
Personnel (regular)	94,200	100,700	100,700	100,700	100,700
Ground forces					
Divisions	4	4	4	4	4
Total number of brigades	14	14	14	14	14
Tanks	~920	~990	~970	~975	921
	(~1,270)	(~1,442)	(~1,467)	(~1,525)	(1,223)*
APCs/AFVs	1,500	1,606	1,813	1,813	1,773
	(1,750)	(1,806)	(2,013)	(2,013)	(2,063)*
Artillery (including MRLs)	838 (863)	838 (863)	844 (867)	844 (867)	871 (896)
Air force					
Combat aircraft	91 (100)	91 (100)	97 (106)	97 (106)	96 (105)
Transport aircraft	12	14	14	14	14
Helicopters	74	74	83	85	83
Air defense forces					
Heavy SAM batteries	14	14	17	17	17
Medium SAM batteries	50	50	50	50	50
Light SAM launchers	50	50	50	50	50
Navy					
Patrol crafts	13	10	10	10	10

* Due to change in estimate

Personnel

	Regular	Reserves	Total
Ground forces	88,000	60,000	148,000
Air force	12,000		12,000
Navy	700		700
Total	**100,700**	**60,000**	**160,700**
Paramilitary			
General Security Forces (including Desert Patrol)	25,000		
Popular Army		200,000-250,000	

Note: The Popular Army is not regarded as a fighting force.

8   KUWAIT

Major Changes

- The American administration approved a Kuwaiti request to acquire Apache AH-64D heli-copters, some of which will be equipped with the Longbow radar.
- The Kuwaiti coast guard received its three new patrol boats from Australia.

General Data

Official Name of the State: State of Kuwait
Head of State: Jabir al-Ahmed al-Jabir al-Sabah
Prime Minister: Sabah al-Ahmed al-Sabah
Minister of Defense: Jabir Mubarak al-Hamad al-Sabah
Chief of General Staff: Major General Fahd Ahmed al-Amir
Commander of the Air Force and Air Defense Forces: Lieutenant General Ibrahim al-Wasmi
Commander of the Navy: Commodore Ahmed Yousuf al-Mualla

Area: 17,820 sq. km. (including 2,590 sq. km. of the Neutral Zone)
Population: 2,600,000

Economic Data (in US $billion)

	2000	2001	2002	2003	2004
GDP (current prices)	37	34.1	35.2	41.6	48.9
Defense expenditure	2.85	3.02	3.54	4.83	NA

Sources: EIU Quarterly Report, EIU Country Profile, IMF International Financial Statistical Yearbook, SIPRI Yearbook

Major Arms Suppliers

Kuwait diversifies its arms procurements. Its major suppliers are the US, the UK, and France. The US supplies attack helicopters, ATGMs, fire control radar, AMARAAM, LASS air defense system, communication systems, maintenance aid, contracting, and upgrading air bases. France supplies heli-copters, MFPBs, SAMs, and anti-ship missiles. The UK supplies APCs, SAMs, and anti-ship missiles.

Other suppliers include Australia (patrol boats), Germany (NBC reconnaissance vehicles, APCs), Jordan (LAVs), PRC (self propelled artillery), and Egypt (air defense missiles).

Foreign Military Cooperation

Type	Details
Foreign forces	Coalition forces: Since the end of the Iraq War there has been a process of force reduction, as some of the forces moved into Iraq and others left the region altogether. As of March 2004, 26,000 US troops remained in Kuwait.
Joint maneuvers	Czech Republic (2003), Egypt (2001), France (2004), GCC countries (2001), Germany (2002), Jordan (2001), UK (2005), US (amphibious, command post, and naval exercises) (2005)
Security agreements	Belarus (2001), France, GCC countries, Iran (2002), Italy (2003), PRC, Russia, South Africa (2002), UK, US

Strategic Assets

NBC Capabilities

Nuclear capability
No known nuclear activity.

Party to the NPT.

Chemical weapons and protective equipment
Fuchs (Fox) ABC detection vehicle (11). Personal protective equipment; unit decontamination equipment.

No known CW activities.

Party to the CWC.

Biological weapons
No known BW activities.

Party to the BWC.

Armed Forces

Order-of-Battle

Year	2001	2002	2003	2004	2005
General data					
Personnel (regular)	19,500	15,500	15,500	15,500	15,500
Ground forces					
Number of brigades	6	7	7	7	7
Tanks	318 (483)	318 (483)	318 (483)	318 (483)	293 (483)
APCs/AFVs	~530	~530	~530	~538	~538
	(797)	(797)	(797)	(805)	(805)
Artillery (including MRLs)	~100	~100	~100	~100	~127
	(~150)	(~130)	(~155)	(~155)	(~155)
Air force					
Combat aircraft	40 (59)	40 (59)	39 (58)	39 (58)	39 (58)
Transport aircraft	5	5	5	5	5
Helicopters	23(28)	25(30)	25(30)	25(30)	25(30)
Air defense forces					
Heavy SAM batteries	12	12	12	12	12
Navy					
Combat vessels	10	10	10	10	10
Patrol craft	69	69	77	77	77

Personnel

	Regular	Reserves	Total
Ground forces	11,000	24,000	35,000
Air force	2,500		2,500
Navy	2,000		2,000
Total	**15,500**	**24,000**	**39,500**
Paramilitary			
National Guard	5,000		5,000
Civil Defense	2,000		2,000

9 LEBANON

Major Changes

- No major changes were recorded in the Lebanese order-of-battle.
- Foreign Syrian and Iranian military forces were withdrawn from Lebanon by May 2005.

General Data

Official Name of the State: Republic of Lebanon
Head of State: President Emile Lahoud
Prime Minister: Fuad Siniora
Minister of Defense: Elias Murr
Commander in Chief of the Armed Forces: Lieutenant General Michel Sulayman
Chief of General Staff: Brigadier General Fady Abu-Shakra
Commander of the Air Force: Brigadier General George Sha ban
Commander of the Navy: Rear Admiral George Ma louf

Area: 10,452 sq. km.
Population: 3,700,000

Economic Data (in US $billion)

	2000	2001	2002	2003	2004
GDP (current prices)	16.4	16.2	17.0	18.0	18.1
Defense expenditure	0.88	0.92	0.81	0.82	NA

Major Arms Suppliers

Lebanon has not received any major arms systems since 1998, when it received helicopters and APCs from the US. Other suppliers are Norway and the Czech Republic, which supplied mine clearing equipment. Iran and Syria supplied artillery rockets, UAVs and other equipment to the Hizbollah militias.

Foreign Military Cooperation

Type	Details
Foreign forces in country	Palestinian organizations; several Iranian instructors with Hizbollah non-government militia in the Beka ; UNIFIL force in south Lebanon (2,000 from France, Ghana, Italy, India, Poland, and Ukraine); Pakistan (300 troops in de-mining operations); UAE (30 army engineers in de-mining operations)

Strategic Assets

NBC Capabilities

Nuclear capability
No nuclear capability.

Party to the NPT.

Chemical weapons and protective equipment
No known CW activity.

Not a party to the CWC.

Biological weapons
No known BW activities.

Party to the BWC.

Armed Forces

Order-of-Battle

Year	2001	2002	2003	2004	2005
General data					
Personnel (regular)	51,400	61,400	61,400	61,400	61,400
Ground forces					
Number of brigades	13	13	13	13	13
Tanks	280 (350)	280 (350)	280 (350)	280 (350)	280 (350)
APCs/AFVs	1,235	1,235	1,235	1,235	1,235
	(1,380)	(1,380)	(1,380)	(1,380)	(1,380)
Artillery (including MRLs)	~335	~335	~335	~335	~335
Air force					
Helicopters	16 (38)	16 (38)	16 (38)	16 (38)	24 (39)*
Navy					
Patrol crafts	32 (35)	20*	20	20	20

* Due to change in estimate
Note: Beginning with this volume of the *Strategic Balance*, the long non-operational winged aircraft of the Lebanese air force have been removed from the table.

Personnel

	Regular	Total
Ground forces	60,000	60,000
Air force	1,000	1,000
Navy	400	400
Total	**61,400**	**61,400**
Paramilitary		
Gendarmerie/ internal security	13,000	

Major Non-Governmental Paramilitary Forces

Personnel

	Active	Reserves	Total
Hizbollah	600 800	3,000 5,000	5,800
Popular Liberation Army (Druze)		10,000	10,000
Amal	50 100	10,000	10,000

Equipment

Hizbollah s equipment includes a large number of artillery rockets, including some heavy, Iranian made Fadjr-3 and Fadjr-5 long-range rockets. They also operate some types of ATGMs and shoulder-launched SAMs. Their aerial power includes some ultra-light aircraft and UAVs.

10   LIBYA

Major Changes

- The Libyan air force absorbed some new transport aircraft acquired from the Ukraine. These include An-124 and An-32. Additional An-32 and An-74 are expected to be delivered soon.

General Data

Official Name of the State: The Great Socialist People s Libyan Arab Jamahiriya
Head of State: Colonel Muammar al-Qaddafi
Prime Minister: Mubarak Abdallah al-Shamikh
Minister of Defense: Colonel Abu-Bakr Yunis Jaber
Inspector General of the Armed Forces: Colonel Mustapha al-Kharrubi
Commander in Chief of the Armed Forces: Colonel Abu-Bakr Yunis Jaber
Chief of Staff: Brigadier General Ahmed Abdallah Awn
Commander of the Air Force and Air Defense Forces: Brigadier General Ali Riffi al-Sharif

Area: 1,759,540 sq. km.
Population: 5,700,000

Economic Data (in US $billion)

	2000	2001	2002	2003	2004
GDP (current prices)	34.3	28.4	19.1	21.2	25.3
Defense expenditure	1.09	1.08	1.37	1.45	NA

Major Arms Suppliers

Libya did not procure major arms systems in the past decade.

Arms suppliers include Ukraine, which supplied transport aircraft and SSMs, and North Korea, which supplied SSMs. Assistance for Libya s weapons program was received mainly from Pakistan, but also from the PRC, Iran, Russia, Ukraine, and private companies in Western Europe.

Major Arms Transfers

Libya sold excess equipment retired from service in the Libyan armed forces to various countries, including Canada (transport aircraft), Pakistan (combat aircraft), UAE (transport aircraft), and Uganda (combat aircraft).

Foreign Military Cooperation

Type	Details
Forces deployed abroad	Since the end of 2001, about 200 Libyan soldiers have been stationed in the Central African Republic (2002)
Security agreements	Algeria (2001), Italy (2003), Tunisia (2001)

Defense Production

Libya produced toxic chemical agents and SSMs. These capabilities were given up in 2003.

Strategic Assets

NBC Capabilities

Nuclear capabilities

5Mw Soviet-made research reactor at Tadjoura; Libya had a clandestine uranium enrichment program with a few thousand centrifuges. These were surrendered and removed in the framework of its steps to renounce its WMD programs.

Party to the NPT. Safeguards agreement with the IAEA in force. Signed but not ratified the African Nuclear Weapon-Free Zone Treaty (Treaty of Pelindaba).

Chemical weapons and protective equipment

Personal protective equipment; Soviet type decontamination units.
CW production facilities, stockpile of chemical agents, nerve gas, and mustard gas.

In the framework of its steps to renounce its WMD programs, work has been carried out to dismantle all past chemical weapons stockpiles. Libya also acceded to the CWC.

Biological weapons

Alleged production of toxins and other biological weapons (unconfirmed).

Party to the BWC.

Ballistic Missiles

Model	Launchers	Missiles	Since	Notes
Scud B	80	500	1976	
Scud C			1999	Scud C missiles have been removed
Total	~80			

Armed Forces

Order-of-Battle

Year	2001	2002	2003	2004	2005
General data					
Personnel	76,000	76,000	76,000	76,000	76,000
SSM launchers	~80	~80	~80	~80	~80
Ground forces					
Number of brigades	1	1	1	1	1
Number of battalions	46	46	46	46	46
Tanks	~650	~650	~650	~650	~650
	(2,210)	(2,210)	(2,210)	(2,210)	(2,210)
APCs/AFVs	~2,750	~2,750	~2,750	~2,750	~2,230
	(2,970)	(2,970)	(2,970)	(2,970)	(2,520)
Artillery (including MRLs)	~2,320	~2,320	~2,320	~2,320	~2,320
	(~2,400)	(~2,400)	(~2,400)	(~2,400)	(~2,400)
Air force					
Combat aircraft	~360 (443)	~340 (443)	~328 (431)	290*(391)	290 (391)
Transport aircraft	85 (90)	85 (90)	68 (73)	68 (73)	72 (77)
Helicopters	127 (204)	127 (204)	112 (189)	112 (189)	112 (189)
Air defense forces					
Heavy SAM batteries	~30	~30	~30	~30	~30
Medium SAM batteries	~10	~10	~10	~10	~10
Light SAM launchers	55	55	55	55	55
Navy					
Submarines	0(4)	0(2)	0(2)	0(2)	0(2)
Combat vessels	24	20	20	23	24

* Due to change in estimate

Personnel

	Regular	Reserves	Total
Ground forces	50,000		50,000
Air force and air defense	18,000		18,000
Navy	8,000		8,000
Total	**76,000**		**76,000**
Paramilitary			
People s Militia	40,000		40,000
Revolutionary Guards	3,000		3,000
(part of the People s Militia)			
Islamic Pan African Legion	2,500		2,500
(part of the People s Militia)			

11 MOROCCO

Major Changes

- No major changes were recorded in the Morocco armed forces.

General Data

Official Name of the State: Kingdom of Morocco
Head of State: King Mohammed VI
Prime Minister: Driss Jettou
Minister of Defense: King Mohammed VI
Secretary General of National Defense Administration: Abdel Rahaman Sbai
Commander in Chief of the Armed Forces: King Mohammed VI
Inspector General of the Armed Forces: General Abdul Aziz Banani
Commander of the Air Force: Ali Abd al-Aziz al-Omrani
Commander of the Navy: Captain Muhammad al-Tariqi

Area: 622,012 sq. km., including the former Spanish Sahara
Population: 30,500,000

Economic Data (in US $billion)

	2000	2001	2002	2003	2004
GDP (current prices)	34.9	36.1	38.3	46.4	52.9
Defense expenditure	1.47	1.5	1.61	NA	NA

Major Arms Suppliers

Most of the Moroccan arsenal comes from France and the US. France supplied patrol boats and helicopters.

Other suppliers include Belarus (tanks), Russia (AD systems), and the UK (upgrading of artillery guns).

Foreign Military Cooperation

Type	Details
Forces deployed abroad	Small contingency force in Bosnia and Croatia
Joint maneuvers	Algeria (2004), France (2004), NATO (2004), Spain (2004), UK (2003), US (2003)
Security agreements	Tunisia, US

Strategic Assets

NBC Capabilities

Nuclear capability
No nuclear capability.

Party to the NPT. Request for nuclear research reactor approved by US government.

Chemical weapons and protective equipment
No known CW activity.

Party to the CWC.

Biological weapons
No known BW activities.

Signed but not ratified the BWC.

Armed Forces

Order-of-Battle

Year	2001	2002	2003	2004	2005
General data					
Personnel (regular)	145,500	145,500	145,500	198,500	198,500
Ground forces					
Number of brigades	6	6	6	6	6
Tanks	640	640	640	640	285 (640)*
APCs/AFVs	1,120 (1,420)	1,120 (1,420)	1,120 (1,420)	1,120 (1,420)	1,089 (1,139)*
Artillery (including MRLs)	1,060	1,060	1,060	1,060	1,060
Air force					
Combat aircraft	59 (72)	59 (72)	59 (72)	59 (72)	59 (72)
Transport aircraft	41 (43)	41 (43)	41 (43)	41 (43)	41 (43)
Helicopters	121 (131)	121 (131)	122 (132)	122 (132)	122 (132)
Air defense forces					
Light SAM launchers	37	37	37	37	37
Navy					
Combat vessels	13	13	15	15	15
Patrol crafts	52	52	52	52	52

* Due to change in estimate
Note: The change in the number of operational MBTs reflects only a change in estimate.

Personnel

	Regular	Reserves	Total
Ground forces	175,000	150,000	325,000
Air force	13,500		13,500
Navy and marines	10,000		10,000
Total	**198,500**	**150,000**	**348,500**
Paramilitary			
Gendarmerie Royale	10,000		10,000
Force Auxiliere	25,000		25,000
Mobile Intervention Corps	5,000		5,000

12 OMAN

Major Changes

- The Omani air force ordered 12 F-16 C/D combat aircraft, scheduled for delivery in 2005. With these advanced aircraft Oman will acquire advanced armament, including PANTERA target acquisition pods, AMRAAM AAMs, Harpoon anti-ship missiles, and JDAM precision guided munitions.
- The air force ordered and received 16 Super Lynx helicopters from Britain.
- The Omani army received the last batch of its VBL light armored vehicles.
- The Omani navy received twelve small fast patrol boats from Abu Dhabi.

General Data

Official Name of the State: Sultanate of Oman
Head of State: Sultan Qabus ibn Said al-Said
Prime Minister: Sultan Qabus ibn Said al-Said
Minister of Defense Affairs: Badr bin Saud bin Harib al-Busaidi
Chief of General Staff: Lieutenant General Ahmed bin Harith bin Naser al-Nabhani
Commander of the Air Force: Air Vice Marshal Yahya bin Rashid al-Juma ah
Commander of the Navy: Rear Admiral Salim bin Abdallah bin Rashid al-Alawi

Area: 212,000 sq. km.
Population: 2,700,000

Economic Data (in US $billion)

	2000	2001	2002	2003	2004
GDP (current prices)	19.86	19.94	20.30	21.60	24.11
Defense expenditure	2.10	2.45	2.53	2.72	NA

Major Arms Suppliers

Major arms suppliers include the US, which supplied combat aircraft, sea launch missiles, AAMs, JDAMs, LGBs, early warning network, and aerial reconnaissance systems; and the UK, which supplied MBTs, missile corvettes, patrol boats, APCs, helicopters and air defense radars.

Other suppliers include UAE (helicopters, assault boats and landing craft), Switzerland (training aircraft), Spain (patrol boats), Pakistan (training aircraft), Italy (helicopters), Netherlands (surveillance radar), PRC (APCs), and France (air defense systems and helicopters).

Foreign Military Cooperation

Type	Details
Foreign forces	Coalition forces: Since the end of the Iraq War there has been a process of force reduction, as some of the forces moved into Iraq and othersleft the region altogether. This process is still currently underway.
Joint maneuvers	Egypt (2001), GCC countries (2003), India (2003), Jordan (2001), Pakistan (2002), UK (2001), US (2001)
Security agreements	GCC countries, India (2003), Iran (2003), Turkey (2001), US, Yemen (2004)

Strategic Assets

NBC Capabilities

Nuclear capability

No known nuclear activity.

Signatory to the NPT.

Chemical weapons and protective equipment

No known CW activities.

Party to the CWC.

Biological weapons

No known BW activities.

Party to the BWC.

Armed Forces

Order-of-Battle

Year	2001	2002	2003	2004	2005
General data					
Personnel (regular)	34,000	34,000	34,000	34,000	34,000
Ground forces					
Number of brigades	4	4	4	4	4
Total number of battalions	18	18	18	18	18
Tanks	151 (201)	151 (201)	151 (201)	151 (201)	124 (201)
APCs/AFVs	~225*	~225	~385	~385	~346
	(~335)	(~335)	(~415)	(~415)	(~476)
Artillery	148 (154)	148 (154)	148 (154)	148 (154)	133 (139)
Air force					
Combat aircraft	29 (30)	29 (30)	29 (30)	29 (30)	29 (30)
Transport aircraft	41 (45)	41 (45)	41 (45)	41 (45)	50 (54)
Helicopters	41	41	46	49	51
Air defense forces					
Light SAM launchers	58	58	58	58	58
Navy					
Combat vessels	9	9	9	9	9
Patrol craft	22	17	37	68	68

*Due to change in estimate

Personnel

	Regular	Reserves	Total
Ground forces	25,000		25,000
Air force	5,000		5,000
Navy	4,000		4,000
Total	**34,000**		**34,000**
Paramilitary			
Tribal force (Firqat)	3,500		3,500
Police/border police (operating aircraft, helicopters and PBs)			7,000
Royal Household (including Royal Guard, Royal Yachts and Royal Flight)			6,500

13   PALESTINIAN AUTHORITY

Major Changes

- After Yasir Arafat s death, Mahmoud Abbas (Abu Mazen) was elected as chairman of the Palestinian Authority. Ahmed Qurei (Abu Ala) retained his post as prime minister.

General Data

This section includes information on the Palestinian Authority and Palestinian security organizations inside Palestinian Authority territory. It does not cover Palestinians living elsewhere.

Official Name: Palestinian National Authority (PA)
Chairman: Mahmoud Abbas (Abu Mazen)
Prime Minister: Ahmed Qurei (Abu Ala)
Minister of Internal security: Nasser Yusef
Chief of Security Forces: Rashid Abu Shback

Area: 400 sq. km. (Gaza), 5,800 sq. km. (West Bank)
Population: Gaza: 1,120,000 est.; West Bank: 2,000,000 est.

Economic Data (in US $billion)

	1999	2000	2001	2002	2003
GDP (current prices)	4.19	4.63	4.03	3.39	NA
Security expenditure	0.50	NA	NA	NA	NA

Major Arms Suppliers

The Palestinian forces smuggle arms from Egypt and Lebanon. Sources of these weapons are not always known.

In 2001 Iran tried to send a shipload of arms, which was intercepted by Israel.

Defense Production

Palestinians forces produce Qassam rockets, mortars, and explosive charges. Palestinian forces announced they managed to produce ATGMs but these are probably unguided rockets.

Security Forces

Order-of-Battle

Year	1999	2000	2001	2002	2003
General data					
Personnel (regular)	~34,000	~36,000	~45,000	~45,000	~45,000
Ground forces					
APCs/AFVs	45	45	~40	+	+
Artillery				+	+
Aerial police					
Helicopters	2 (4)	2 (4)	2 (4)	0	0
Coastal police					
Patrol craft	13	10	13	0	0

Military Forces

Personnel

	Gaza	West Bank	Total	Notes
General Security Service branches				
Public security	+	+	+	Also referred to as the National Security Force
Coastal police	+	+	+	
Aerial police	+	+	+	Rudimentary unit operating VIP helicopters
Civil police	+	+	+	Civilian police a law enforcement agency; operates the 700-strong rapid deployment special police
Preventive Security Force	+	+	+	Plainclothes internal security force
General Intelligence	+	+	+	Intelligence gathering organization
Military Intelligence	+	+	+	Unrecognized preventive security force; includes the Military Police
Civil Defense	+	+	+	Emergency and rescue service
Additional security forces				
Presidential Security	+	+	+	Elite unit responsible for security
Special Security Force	+	+	+	Unrecognized intelligence organization
Total	**~25,000**	**~20,000**	**+**	

Note: More than four years of armed conflict between the PA and Israel changed the situation described by this table. Most of the fighting was done by the unofficial organizations like Tanzim and Hamas. At present there is no data concerning the status of the organizations listed above. Some of them ceased to exist, and some may reappear as strong organizations, depending on the personal status of their leaders. Thus, we prefer to display statistics as they were at the end of 2000. The Palestinian security services included several organizations under the Palestinian Directorate of Police Force recognized in the Cairo and Washington agreements. Some of the security organizations (particularly the civilian police) have little or no military significance. They are mentioned here because of the unusual organizational structure, and because it is difficult to estimate the size of the total forces that do have military significance. Since the death of Yasir Arafat there have been attempts to consolidate the various organizations. As of now the situation is still in constant flux.

14   QATAR

Major Changes

- No major changes were recorded for the Qatari armed forces.

General Data

Official Name of the State: State of Qatar
Head of State: Shaykh Hamad ibn Khalifa al-Thani
Prime Minister: Abdallah Ibn Khalifa al-Thani
Minister of Defense: Shaykh Hamad ibn Khalifa al-Thani
Commander in Chief of the Armed Forces: Shaykh Hamad ibn Khalifa al-Thani
Chief of General Staff: Brigadier General Hamad bin Ali al-Attiyah
Commander of the Ground Forces: Colonel Saif Ali al-Hajiri
Commander of the Air Force: General Ali Saeed al-Hawal al-Marri
Commander of the Navy: Captain Said al-Suwaydi

Area: 11,437 sq. km.
Population: 600,000

Economic Data (in US $billion)

	2000	2001	2002	2003	2004
GDP (current prices)	17.8	17.7	17.9	19.5	21.9
Defense expenditure	NA	NA	NA	NA	NA

Major Arms Suppliers

France is the major arms supplier to Qatar. It supplied patrol boats.
The US built major installations, including a large air base and storage facilities. These installations are currently used by US forces.

Foreign Military Cooperation

Type	Details
Forces deployed abroad	Troops part of GCC Peninsula Shield rapid deployment force in Saudi Arabia
Foreign forces in country	Coalition forces: Since the end of the Iraq War there has been a process of force reduction, as some of the forces moved into Iraq and others left the region altogether. This process is still underway.
Joint maneuvers	France (2002), GCC countries (2003), Italy, UK, US (2002), Yemen
Security agreements	Bahrain, France (2004), GCC defense pact (2000), Italy (2001), Iran, Kuwait, Saudi Arabia, US (2003)

Strategic Assets

NBC Capabilities

Nuclear capability

No known nuclear activity.

Party to the NPT.

Chemical weapons and protective equipment

No known CW activities.

Party to the CWC.

Biological weapons

No known BW activities.

Party to the BWC.

Armed Forces

Order-of-Battle

Year	2001	2002	2003	2004	2005
General data					
Personnel (regular)	11,800	11,800	11,800	11,800	11,800
Ground forces					
Number of brigades	2	2	2	2	2
Number of regiments					
Total number of battalions	11	11	11	11	11
Tanks	44	44	44	44	30*
APCs/AFVs	~260 (338)	~260 (338)	~260 (338)	~260 (338)	~280 (310)*
Artillery (including MRLs)	56	56	56	56	56
Air force					
Combat aircraft	18	18	18	18	12*
Transport aircraft	7 (8)	7 (8)	7 (8)	7 (8)	7 (8)
Helicopters	30 (31)	30 (31)	30 (31)	30 (31)	24 (25)*
Air defense forces					
Light SAM launchers	51	51	51	51	51
Navy					
Combat vessels	7	7	7	7	7
Patrol crafts	26	13	13	13	13

* Due to change in estimate

Personnel

	Regular	Total
Ground Forces	8,500	8,500
Air Force	1,500	1,500
Navy (including marine police)	1,800	1,800
Total	**11,800**	**11,800**
Paramilitary		
Armed police	8,000	8,000

15 SAUDI ARABIA

Major Changes

- The Saudi air force introduced its new Super Mushshak light training aircraft.
- No other major changes were recorded for the Saudi Arabian armed forces.

General Data

Official Name of the State: The Kingdom of Saudi Arabia
Head of State: King Abdullah ibn Abd al-Aziz al-Saud
Prime Minister: King Abdullah ibn Abd al-Aziz al-Saud
First Deputy Prime Minister and Heir Apparent: Crown Prince Sultan ibn Abd al-Aziz al-Saud
Defense and Aviation Minister: Crown Prince Sultan ibn Abd al-Aziz al-Saud
Chief of General Staff: General Salih ibn Ali al-Muhaya
Commander of the Ground Forces: Lieutenant General Husein al-Qubeel
Commander of the National Guard: Crown Prince Sultan ibn Abd al-Aziz al-Saud
Commander of the Air Force: Lieutenant General Abd al-Rahman ibn Fahd al-Faisal
Commander of the Navy: Vice Admiral Fahd ibn Abdullah

Area: 2,331,000 sq. km.
Population: 25,200,000

Economic Data (in US $billion)

	2000	2001	2002	2003	2004
GDP (current prices)	188.7	183.3	188.8	214.7	241.0
Defense expenditure	20	21.14	18.7	19.1	NA

Major Arms Suppliers

The major arms suppliers to Saudi Arabia are the US and France. The US supplied combat aircraft, AAMs, surveillance radars, anti-tank missiles, and early warning networks. France supplied combat vessels, upgrading of SAMs, and C⁴I systems.

Other suppliers include Italy (helicopters and naval guns), Jordan (RVs), UK (aircraft and hovercraft), and Canada (APCs, IFVs).

Foreign Military Cooperation

Type	Details
Foreign forces	GCC Peninsula Shield rapid deployment force: 7,000-10,000 men at Hafr al-Batin, mostly Saudis, and from other GCC countries; some 8,000 instructors and technicians from Pakistan; UK (25) (2004). The US has withdrawn most of its military presence in Saudi Arabia. Current American presence includes some 950 personnel.
Joint maneuvers	Egypt (2004), France, GCC countries (2004), Jordan (2001), Pakistan (2004), UK, US (2001), Yemen (2005)

Defense Production

APCs, radar subsystems, parts of EW equipment.

Strategic Assets

NBC Capabilities

Nuclear capability
No known nuclear activity.

Party to the NPT.

Chemical weapons and protective equipment
Personal protective equipment; decontamination units; US-made CAM chemical detection systems; Fuchs (Fox) NBC detection vehicles.

No known CW activities.

Party to the CWC.

Biological weapons
No known BW activities.

Party to the BWC.

Ballistic Missiles

Model	Launchers	Missiles	Since	Notes
CSS-2	8-12	30-50	1988	Number of launchers unconfirmed

Space Assets

Model	Type	Notes
Satellites		
Arabsat	Communication	Civilian
Saudi Comsat 1/2	Communication	Commercial; 12 kg. each
SaudiSat 1A/1B/1C/2	Remote sensing and space research	2 (10 kg. each) were launched in September 2000 by a Russian military rocket, and are orbiting 650 km above earth. The third satellite was launched in December 2002. Saudi Sat 2 (30kg.) was launched in June 2004
Ground Stations		
SCRS	Imagery	Receiving SPOT, Landsat, and NOAA

Armed Forces

Order-of-Battle

Year	2001	2002	2003	2004	2005
General data					
Personnel (regular)	171,500	171,500	171,500	171,500	171,500
SSM launchers	8-12	8-12	8-12	8-12	8-12
Ground forces					
Number of brigades	20	20	20	20	20
Tanks	750	750	750	750	750
	(1,015)	(1,015)	(1,015)	(1,015)	(1,015)
APCs/AFVs	~4,500	~4,630	~4,630	~4,630	~4,430
	(~5,300)	(~5,430)	(~5,430)	(~5,430)	(~5,230)
Artillery	~410	~410	~410	~410	~410
(incl. MRLs)	(~780)	(~780)	(~780)	(~780)	(~780)
Air force					
Combat aircraft	~360	~345	~289 *	286*	256*
	(~365)		(~344)	(~340)	(~340)
Transport aircraft	42 (55)	42 (55)	42 (55)	42 (55)	38 (51)
Helicopters	214 (216)	214 (216)	214 (216)	214 (216)	214 (216)
Air defense forces					
Heavy SAM batteries	25	25	25	25	25
Medium SAM batteries	21	21	21	21	21
Navy					
Combat vessels	24	25	27	27	27
Patrol craft	74	64	68	68	68

* Due to change in estimate

Personnel

	Regular	Reserves	Total
Ground forces	**75,000**		**75,000**
Air force	20,000		20,000
Air defense	4,000		4,000
Navy (including a marine unit)	13,500		13,500
National Guard	57,000	20,000	77,000
Royal Guard	2,000		2,000
Total	**171,500**	**20,000**	**191,500**
Paramilitary			
Mujahidun (affiliated with National Guard)		30,000	30,000
Coast Guard	4,500		4,500
Frontier Corps	10,500		10,500

16 SUDAN

Major Changes

- No major changes were recorded for the Sudanese armed forces.

General Data

Official Name of the State: The Republic of Sudan
Head of State: President Omar Hassan Ahmed al-Bashir
Defense Minister: Major General Bakri Hassan Sallah
Chief of General Staff: General Abbas Arabi
Commander of the Air Force: Major General Ali Mahjoub Mardi
Commander of the Navy: Commodore Abbas al-Said Othman

Area: 2,504,530 sq. km.
Population: 34,300,000

Economic Data (in US $billion)

	2000	2001	2002	2003	2004
GDP (current prices)	11.2	12.1	13.6	15.6	19.2
Defense expenditure	0.32	NA	0.35	0.32	NA

Major Arms Suppliers

Major arms sales to Sudan come from Russia, which supplied combat aircraft and APCs. Previously Sudan received arms from Iran, which allegedly supplied tanks, aircraft, other vehicles, and EW equipment.

Other suppliers included Belarus (combat aircraft), India (surveillance radar), Lithuania (helicopters), PRC (upgrading of aircraft), Poland (tanks), and Ukraine (aircraft).

Foreign Military Cooperation

Type	Details
Forces deployed abroad	Central African Republic (2002)
Foreign forces	PRC (alleged presence of forces for the defense of Chinese-operated oil fields) (2001); African Union protection force in Darfur province (3200 personnel)
Security agreements	Syria (2000), Egypt (2001), Russia (2002)

Strategic Assets

NBC Capabilities

Nuclear capability
No known nuclear activity.

Party to the NPT.

Chemical weapons and protective equipment
Alleged CW from Iran (unsubstantiated); alleged production of CW (unsubstantiated); personal protective equipment; unit decontamination equipment.

Party to the CWC.

Biological weapons
No known BW activities.

Party to the BWC.

Armed Forces

Order-of-Battle

Year	2001	2002	2003	2004	2005
General data					
Personnel (regular)	103,000	104,000	104,000	104,000	104,000
Ground forces					
Divisions	9	9	9	9	9
Total number of brigades	61	61	61	61	61
Tanks	~350	~350	~350	~350	~350
APCs/AFVs	~560	~545	~575	~575	~575
	(~700)	(~745)	(~745)	(~745)	(~745)
Artillery (including MRLs)	~760	~770	~778	~778	~778
	(~770)	(~785)	(~793)	(~793)	(~793)
Air force					
Combat aircraft	~35 (55)	~35 (55)	~51 (71)	~50 (72)	~40 (62)
Transport aircraft	24	24	24	24	14
Helicopters	57 (73)	59 (71)	75 (87)	75 (87)	46 (51)
Air defense forces					
Heavy SAM batteries	20	20	20	20	20
Navy					
Patrol craft	18	18	16	16	16

Personnel

	Regular	Reserves	Total
Ground forces	100,000		100,000
Air force	3,000		3,000
Navy	1,000		1,000
Total	**104,000**		**104,000**
Paramilitary			
People s Defense Forces	15,000	85,000	100,000
Border Guard	2,500		2,500

17 SYRIA

Major Changes

* No major changes were recorded for the Syrian armed forces.
* Syria withdrew its military forces from Lebanon by May 2005.

General Data

Official Name of the State: The Arab Republic of Syria
Head of State: President Bashar al-Asad
Prime Minister: Mohammed Jazi Otri
Minister of Defense: Major General Hassan Turkamani
Chief of General Staff: Major General Ali Habib
Commander of the Air Force: Major General Hazem Khadra a
Commander of the Navy: Vice Admiral Wa il Nasser

Area: 185,180 sq. km.
Population: 18,200,000

Economic Data (in US $billion)

	2000	2001	2002	2003	2004
GDP (current prices)	19.5	20.6	21.6	22.5	22.3
Defense expenditure	1.07	1.33	1.35	1.6	NA

Major Arms Suppliers

Russia was the major supplier, and it remains the major potential supplier. In the past it supplied Syria with all its major armament systems. Recent arms deals included ATGMs and AD systems.

Other suppliers include Iran (ballistic missile technology), North Korea (ballistic missiles), PRC (ballistic missiles), Belarus (AD systems), Ukraine (upgrading of tanks, radars), Armenia (upgrading of tanks), Bulgaria (upgrading of APCs), and Italy (upgrading of tanks).

Major Arms Transfers

Lebanon (artillery rockets).

Foreign Military Cooperation

Type	Details
Forces deployed abroad	15,000 in Beka , northern Lebanon (Tripoli area), and Beirut

Defense Production

Ballistic missiles, artillery rockets, upgrading of tanks.

Strategic Assets

NBC Capabilities

Nuclear capability
Basic research. Alleged deal with Russia for a 24 Mw reactor. Deals with China for a 27 kw reactor and with Argentina for a 3 Mw research reactor are probably cancelled. Party to the NPT. Safeguards agreement with the IAEA in force.

Chemical weapons and protective equipment
Stockpiles of nerve gas, including sarin, mustard, and VX.

There are unconfirmed allegations that Syria received Iraq s stockpile of chemical weapons just before the Iraq War broke out.

Delivery vehicles include chemical warheads for SSMs and aerial bombs.

Personal protective equipment; Soviet-type unit decontamination equipment.

Not a party to the CWC.

Biological weapons
Biological weapons and toxins (unconfirmed).

Signed but not ratified the BWC.

Ballistic Missiles

Model	Launchers	Missiles	Since	Notes
SS-1 (Scud B)	18	200	1974	
SS-1 (Scud C)	8	80	1992	
SS-21 (Scarab)	18		1983	
Scud D	+		2002	
Total	~45			

Space Assets

Name	Type	Notes
Satellite imaging GORS	Remote sensing	Using images from Cosmos, ERS, Landsat, SPOT satellites

Armed Forces

Order-of-Battle

Year	2001	2002	2003	2004	2005
General data					
Personnel (regular)	380,000	380,000	289,000*	289,000	289,000
SSM launchers	44	~45	~45	~45	~45
Ground forces					
Divisions	12	12	12	12	12
Total number of brigades	67	67	67	67	67
Tanks	3,700	3,700	3,700	3,700	3,700
	(4,800)	(4,800)	(4,800)	(4,800)	(4,800)
APCs/AFVs	~5,000	5,060	5,060	5,060	5,060
Artillery (including MRLs)	~2,600	~2,600	~2,600	3,000	3,274
	(~3,000)	(~3,000)	(~3,000)	(3,400)	(3,674)*
Air force					
Combat aircraft	490	490	450* (490)	450 (490)	350 (490)*
Transport aircraft	23 (25)	23	23	23	23
Helicopters	285	225*	225	213 (225)	195 (225)*
Air defense forces					
Heavy SAM batteries	108	108	108	108	108
Medium SAM batteries	64	64	64	64	64
Light SAM launchers	55	55	55	55	55
Navy					
Submarines	0 (3)				
Combat vessels	14	14	16*	16	16
Patrol craft	8	8	8	8	8

* Due to change in estimate

Personnel

	Regular	Reserves	Total
Ground forces	215,000	100,000	315,000
Air force	30,000	10,000	40,000
Air defense	40,000	20,000	60,000
Navy	4,000	2,500	6,500
Total	**289,000**	**132,500**	**421,500**
Paramilitary			
Gendarmerie	8,000		8,000
Workers Militia		400,000	400,000

18   TUNISIA

Major Changes

- The Tunisian navy received six Type 142 combat vessels from Germany.
- No other major changes were recorded for the Tunisian armed forces.

General Data

Official Name of the State: The Republic of Tunisia
Head of State: President Zayn al-Abedine Bin Ali
Prime Minister: Mohamed Ghannouchi
Minister of Defense: Dali Jazi
Secretary of State for National Defense: Chokri Ayachi
Commander of the Ground Forces: Brigadier General Rashid Amar
Commander of the Air Force: Major General Rida Hamuda Atar
Commander of the Navy: Commodore Brahim Barak

Area: 164,206 sq. km.
Population: 9,900,000

Economic Data (in US $billion)

	2000	2001	2002	2003	2004
GDP (current prices)	19.4	20.0	21.1	25.0	28.3
Defense expenditure	0.32	0.32	NA	NA	NA

Major Arms Suppliers

Tunisia had no major arms deals in the past decade. Minor acquisitions were from France (APCs), Germany (MFPBs), Spain (patrol boats), US (transport aircraft), and Italy (transport aircraft).

Foreign Military Cooperation

Type	Details
Forces deployed abroad	Burundi (ONUB), Congo (MONUC) , Ethiopia and Eritrea (UNMEE), Ivory Coast (UNOCI)
Joint maneuvers	Algeria (2004), France, Greece (2003), Spain (unconfirmed), US
Security cooperation	Egypt , Germany (2003), Greece, Morocco

Defense Production

Patrol boats.

Strategic Assets

NBC Capabilities

Nuclear capability
No known nuclear activity.

Signatory to the NPT.

Chemical weapons and protective equipment
No known CW activities.

Party to the CWC.

Biological weapons
No known BW activities.

Party to the BWC.

Armed Forces

Order-of-Battle

Year	2001	2002	2003	2004	2005
General data					
Personnel (regular)	35,500	35,500	35,500	35,500	35,500
Ground forces					
Number of brigades	5	5	5	5	5
Tanks	139 (144)	139 (144)	139 (144)	139 (144)	139 (144)
APCs/AFVs	316	326	326	326	326
Artillery (including MRLs)	205 (215)	205	205	205	205
Air force					
Combat aircraft	18	18	18	18	18
Transport aircraft	9 (11)	15 (17)	15 (17)	15 (17)	15 (17)
Helicopters	51	49	47	47	47
Air defense forces					
Light SAM launchers	83	83	83	83	83
Navy					
Combat vessels	9	9	9	15	15
Patrol crafts	37	35	36	40	40

Personnel

	Regular	Reserve	Total
Ground forces	27,000		27,000
Air force	4,000		4,000
Navy	4,500		4,500
Total	**35,500**		**35,500**
Paramilitary			
Gendarmerie	2,000		2,000
National Guard	7,000		7,000

19 TURKEY

Major Changes

- The Turkish armed forces received their first indigenously made light air defense systems   the Atilgan and the Zipkin.
- An ongoing project is the renewal of the Turkish artillery with 300 Firtina 155mm self-propelled guns and 400 Panther 155mm towed howitzers, all indigenously assembled. These weapon systems began entering service.
- The air force launched a project to upgrade all of its 218 F-16 combat aircraft.
- The navy and the coast guard received all of their nine CN-235 maritime patrol aircraft, and ordered additional 10 ATR-72 patrol aircraft. These too would use the same mission hardware by Thales.

General Data

Official Name of the State: Republic of Turkey
Head of State: President Ahmet Necdet Sezer
Prime Minister: Recep Tayyip Erdogan
Minister of National Defense: Vecdi Gonul
Chief of General Staff: General Hilmi Ozkok
Commander of the Ground Forces: General Yasar Buyukanit
Commander of the Air Force: General Ibrahim Firtina
Commander of the Navy: Admiral Ozden Ornek

Area: 780,580 sq. km.
Population: 72,300,000

Economic Data (in US $billion)

	2000	2001	2002	2003	2004
GDP (current prices)	199.3	145.6	184.2	239.7	293.4
Defense expenditure	10.0	7.21	9.05	11.6	NA

Major Arms Suppliers

Turkey s major arms suppliers are the US, France, and Germany. The US supplied combat vessels, combat aircraft, helicopters, early warning aircraft, AD missiles, anti-tank missiles, and radars. France supplied combat vessels, helicopters, training aircraft, and cruise missiles. Germany supplied submarines, combat vessels, and radars.

Other suppliers include Israel (upgrading of combat aircraft, upgrading of tanks, upgrading of heli-copters, anti-radiation drones, cruise missiles, and reconnaissance pods), South Korea (self-propelled artillery guns), Italy (helicopters, radars, maritime patrol aircraft), Norway (cruise missiles), Spain (transport aircraft), and the UK (AD missiles).

Major Arms Transfers

Turkey sold armament systems to several countries, including Malaysia (IFVs), UAE (IFVs), Israel (APCs), Jordan (transport aircraft, APCs), Azerbaijan (patrol boats, APCs), Georgia (patrol boats, helicopters), Kazakhstan (patrol boats, APCs), and Macedonia (combat aircraft).

Foreign Military Cooperation

Type	Details
Forces deployed abroad	Afghanistan (1,800 troops in ISAF); Albania; Bosnia (350 troops); Cyprus (30,000 troops); Georgia (UNOMIG); northern Iraq (1,000 troops); Israel (TIPH), Ivory Coast (UNOCI); Kosovo (UNMIK); Liberia (UNMIL); Macedonia (140 troops)
Foreign forces	US (1,760 troops as of September 2004)
Cooperation in training	Albania, Azerbaijan, Israel (mutual use of airspace and training facilities), Jordan (mutual use of airspace and training facilities; joint training of infantry), Georgia, PRC
Joint maneuvers	Albania (naval 2000), Bulgaria (2003), Georgia (naval), Germany (2003), Israel (2003), Jordan, Macedonia (part of multinational peace keeping brigade 2000), NATO member states (2000), Netherlands (2004), Pakistan, Poland, Romania (part of multinational peace-keeping brigade 2000), US (2003)
Security agreements	Bosnia (2003), Chile (2004), Croatia (2001), France, Georgia (2001), Kazakhstan (2000), Latvia, Mauritania (2005), Oman (2001), Pakistan (2003), Syria (2002)

Defense Production

Turkey has a large and diversified defense industry. Its aerospace industry produces or assembles combat aircraft, helicopters, and transport and training aircraft. Its naval industry produces or assembles submarines, combat vessels, and patrol boats. Land based systems produced include ACVs, self-propelled guns, and AD systems. Other munitions produced include artillery rockets and anti-tank missiles. The electronic industry produces radars, electronic warfare systems, and fire control systems.

Strategic Assets

NBC Capabilities

Nuclear capability

One 5Mw TR-2 research reactor at Cekmerce and one 250 kw ITV-TRR research reactor at Istanbul. Turkey intends to order a 1,000 Mw reactor. As a member of NATO, nuclear weapons were deployed in Turkey in the past, and might be deployed there again.

Party to the NPT. Safeguards agreement with the IAEA in force.

Chemical weapons and protective equipment

Personal protective suits; portable chemical detectors; Fox detection vehicles.

Party to the CWC.

Biological weapons

No known BW activity.

Party to the BWC.

Ballistic Missiles

Model	Launchers	Missiles	Since	Notes
ATACMS	12	72	1997	Using MLRS launchers
Future procurement				
J project			2001	Under development

Space Assets

Model	Type	Notes
Ground stations		
BILTEN	Remote sensing	Receiving imagery from Bilsat
SAGRES	Remote sensing	Receiving imagery from SPOT, ERS, RADARSAT, and NOAA
Satellites		
Turksat-2A	Communication	Both civilian and military
Bilsat	Remote sensing	120 kg payload, 686 km orbit, 12m resolution earth observation civilian satellite
Satellite imagery		
Ikonos	Reconnaissance	Commercial satellite imagery
Ofeq 5	Reconnaissance	Sharing of satellite imagery by Israel

Armed Forces

Order-of-Battle

Year	2001	2002	2003	2004	2005
General data					
Personnel (regular)	610,000	515,000	515,000	421,000	421,000
SSM launchers	12	12	12	12	12
Ground forces					
Divisions	5	3	3	3	3
Total number of brigades	67	63	63	59	59
Tanks	2,600	2,600	2,600	2,600	2,600
	(4,255)	(4,255)	(4,255)	(4,255)	(4,180)
APCs/AFVs	5,460	5,460	5,790	5,790	5,885
Artillery (including MRLs)	~4,350	~4,355	~4,370	~4,370	~4,370
	(~4,650)	(~4,655)	(~4,670)	(~4,670)	(~4,670)
Air force					
Combat aircraft	~445 (465)	~390 (410)	~400 (422)	~400 (420)	~400 (420)
Transport aircraft	90 (94)	93 (97)	90 (100)	90 (100)	90 (107)
Helicopters	407	461	462	462	467
Air defense forces					
Heavy SAM batteries	24	24	24	24	24
Light SAM launchers	86	86	86	86	96
Navy					
Submarines	14	13	12	12	12
Combat vessels	78	83	84	84	83
Patrol crafts	103	113	102	105	106

Personnel

	Regular	Reserves	Total
Ground forces	308,000	259,000	567,000
Air force	60,000	65,000	125,000
Navy	53,000	55,000	108,000
Total	**421,000**	**379,000**	**800,000**
Paramilitary			
Coast Guard	2,200		2,200
Gendarmerie/			
National Guard	180,000	50,000	230,000

20 UNITED ARAB EMIRATES (UAE)

Major Changes

- Delivery of Leclerc MBTs and the other vehicles ordered within the framework of the same contract was concluded.
- The army received its first Guardian ACVs from Ukraine. The turrets for these ACVs were ordered separately and will be installed indigenously. Similar turrets will be installed on some of the older Scorpion light tanks.
- The Emiri air force is scheduled to receive its first F-16E/F during 2005.
- Meanwhile the Emiri air force received most of its Mirage 2000-9. The delivery of all 63 aircraft (new and refurbished) is scheduled to be concluded in 2005.
- The Emiri navy is absorbing its 24 meter Transportbat 2000 landing craft (out of twelve ordered under the Ghannata project). The navy also ordered three larger (64 m.) landing craft, and is continuing the process of upgrading its TNC-45 MFPBs.
- The Emiri coast guard is absorbing its second batch of 30 small, locally produced, fast attack boats, in addition to 24 vessels which are already in service.

General Data

Official Name of the State: United Arab Emirates
Head of State: Shaykh Zayid ibn Sultan al-Nuhayan, Emir of Abu Dhabi
Prime Minister: Shaykh Maktum ibn Rashid al-Maktum, Emir of Dubai
Minister of Defense: Muhammad ibn Rashid al-Maktum
Chief of General Staff: HRH Lieutenant General Hamad Muhammad Thani al-Rumaithi
Commander of the Air Force and
Air Defense Forces: Brigadir General Khalid bin Abdullah al-Buainnain
Commander of the Navy: Brigadir General Suhail Shaheen al-Murar

Area:. 82,900 sq. km. est.
Population: 4,300,000 est.
Note: The UAE consists of seven principalities: Abu Dhabi, Dubai, Ras al-Khaima, Sharja, Umm al-Qaiwain, Fujaira, and Ajman

Economic Data (in US $billion)

	2000	2001	2002	2003	2004
GDP (current prices)	70.2	69.2	71.2	79.8	90.0
Defense expenditure	2.36	2.35	2.46	2.55	NA

Major Arms Suppliers

The US and France are UAE s major arms suppliers. The US supplied combat aircraft, attack heli-copters, AAMs, JDAMs, naval SAMs, anti-tank missiles, anti-ship missiles, command and control aircraft, and advanced air launched munitions. France supplied combat aircraft, helicopters, MBTs, MFPS, SAMs, ARVs, LAVs, torpedoes, anti-ship missiles, and C³I systems.

Other suppliers include Belgium (upgrading of AVs), Denmark (naval surveillance radar), Netherlands (frigates), Germany (APCs, tank transporters, ABC detection vehicles, naval systems), Italy (procure-ment and upgrading of helicopters, naval systems), Jordan (RVs), Libya (helicopters), Romania (upgrading of helicopters), Russia (procurement and upgrading of IFVs, air defense systems), Spain (patrol aircraft), Sweden (naval systems), Switzerland (APCs), UK (laser pointing systems), Ukraine (APCs), and South Africa (EW systems, UAVs).

Major Arms Transfers

The UAE sold armament systems to several countries, including Iraq (aircraft and APCs), Jordan (upgrading of light tanks), Oman (assault boats and landing craft), and Yemen (patrol boats).

Foreign Military Cooperation

Type	Details
Forces deployed abroad	Saudi Arabia (part of GCC Peninsula Shield rapid deployment force)
Foreign forces	Some 150 US soldiers (2004)
Joint maneuvers	Egypt (2001), France (2005), GCC countries (2002), Jordan (2001), Turkey (2002), US (2004)
Security agreements	France (2000), Germany, India (2003), Yemen (2005)

Defense Production

UAE s industry produces patrol boats, corvettes, amphibious landing craft, and assembles UAVs and mini-UAVs, target drones, and various reconnaissance vehicles.

Strategic Assets

NBC Capabilities

Nuclear capability
No known nuclear activity.

Signatory to the NPT.

Chemical weapons and protective equipment
No known CW activities. Personal protective equipment; unit decontamination equipment.

Signed and ratified the CWC.

Biological weapons
No known BW activities.

Signed but not ratified the BWC.

Ballistic Missiles

Model	Launchers	Missiles	Since	Notes
Scud B	6		1991	Owned by Dubai; unconfirmed

Space Assets

Model	Type	Notes
Satellites Thuraya-1/2	Communication	Geosynchronous, civilian satellite The first was launched in September 2000, the second in June 2003
Ground stations Dubai Space Imaging	Remote sensing	Receiving satellite images from Ikonos and India s IRS satellites

Armed Forces

Order-of-Battle

Year	2001	2002	2003	2004	2005
General data					
Personnel (regular)	46,500	65,500	65,500	65,500	65,500
SSM launchers	6	6	6	6	6
Ground forces					
Number of brigades	8	8	8	8	8
Tanks	~400 (~470)	532 (604)	539 (611)	532 (604)	532 (604)
APCs/AFVs	~1,250 (~1,410)	~1,190 (~1,350)	~1,190 (~1,350)	~1,190 (~1,350)	~1,044
Artillery (including MRLs)	399 (422)	399 (422)	405 (428)	405 (428)	337*(360)
Air force					
Combat aircraft	54 (66)	54 (66)	47 (70)	47 (70)	93 (103)
Transport aircraft	33 (36)	33 (36)	33 (36)	33 (36)	30 (33)
Helicopters	91 (103)	102 (114)	102 (114)	102 (114)	102 (109)
Air defense forces					
Heavy SAM batteries	5	5	5	5	5
Medium SAM batteries	6	6	6	6	6
Light SAM launchers	~115	~115	~115	~115	~115
Navy					
Combat vessels	12	12	12	12	12
Patrol craft	112	112	128	134	134

* Due to changes in estimate

Personnel

	Regular	Reserves	Total
Ground forces	59,000		59,000
Air force	4,500		4,500
Navy	2,000		2,000
Total	**65,500**		**65,500**
Paramilitary			
Coast Guard	+		+
Frontier Corps	+		+

21 YEMEN

Major Changes

*   Yemen continues working to improve its defense relations with the US. The US is training Yemeni personnel in counterterrorism activities and helping Yemen to establish its coast guard. It also supplies Yemen with spare parts for aircraft.
*   The Yemeni air force received its first upgraded MiG-29 combat aircraft. The delivery of all 16 aircraft is scheduled to be concluded by the end of 2005.
*   The Yemeni navy received its ten Bay class fast patrol boats from Australia.

General Data

Official Name of the State: Republic of Yemen
Head of State: President Ali Abdallah Salih
Prime Minister: Abd al-Qadir Ba Jamal
Minister of Defense: Brig. General Abdallah Ali Alaywa
Chief of General Staff: Brig. General Abdallah Ali Alaywah
Commander of the Air Force: Colonel Muhammad Salih al-Ahmar
Commander of the Navy: Admiral Abdallah al-Mujawar

Area: 527,970 sq. km.
Population: 20,700,000

Economic Data (in US $billion)

	2000	2001	2002	2003	2004
GDP (current prices)	9.5	9.6	10.3	11.3	13.7
Defense expenditure	0.44	0.47	0.60	NA	NA

Major Arms Suppliers

Russia is Yemen s major arms supplier. It supplied combat aircraft, helicopters, and MBTs.

Other suppliers include Australia (patrol boats), France (military boats), North Korea (SSMs), UAE (patrol boats), Poland (landing craft), Czech Republic (training aircraft, tanks), and the US (C^3 systems, patrol boats).

Foreign Military Cooperation

Type	Details
Forces deployed abroad	Burundi (ONUB); Liberia (NUMIL)
Foreign forces	Some 20 US soldiers (2004)
Security agreements	France (2004); Oman (2004); Turkey (2002); UAE (2005)
Joint maneuvers	Saudi Arabia (2005); US (2004)

Strategic Assets

NBC Capabilities

Nuclear capability
No known nuclear activity.

Signatory to the NPT.

Chemical weapons and protective equipment
No known CW activities.

Signed and ratified the CWC.

Biological weapons
No known BW activities.

Party to the BWC.

Ballistic Missiles

Model	Launchers	Missiles	Since	Notes
SS-1 (Scud B)	6			New missiles received from North Korea, possibly Scud C
SS-21 (Scarab)	4		1988	
Total	10			

Note: Serviceability of missiles and launchers unknown.

Armed Forces

Note: Since the 1994 civil war, all figures are rough estimates.

Order-of-Battle

Year	2001	2002	2003	2004	2005
General data					
Personnel (regular)	~65,000	~65,000	~65,000	~65,000	~65,000
SSM launchers	10	10	10	10	10
Ground forces					
Number of brigades	33	33	33	33	33
Tanks	~715	~715	~715	~715	~715
	(~1,180)	(~1,180)	(~1,180)	(~1,180)	(~1,180)
APCs/AFVs	~480	~495	~495	~495	~495
	(~1,200)	(~1,210)	(~1,210)	(~1,210)	(~1,140)
Artillery	~670	~675	~675	~675	~675
(including MRLs)	(~1,000)	(~1,025)	(~1,025)	(~1,025)	(~995)
Air force					
Combat aircraft	~55	~65	~65	~54	52
	(~180)	(~190)	(~190)	(~175)	(171)
Transport aircraft	20 (30)	20 (30)	20 (30)	20 (30)	14
Helicopters	26 (70)	26 (70)	26 (70)	26 (70)	24 (68)
Air defense forces					
Heavy SAM batteries	25	25	25	25	25
Medium SAM batteries	+	+	+	+	+
Light SAM launchers	120	120	120	120	120
Navy					
Combat vessels	10	10	10	10	10
Patrol craft	9	8	8	8	8

Personnel

	Regular	Reserves	Total
Ground forces	~60,000	200,000	~260,000
Air force	3,000		3,000
Navy	2,000		2,000
Total	**~65,000**	**200,000**	**~265,000**
Paramilitary			
Central Security Force	50,000		50,000

Tables and Charts

The Middle East Military Balance at a Glance

State	Personnel			Ground Forces			
	Regular	Reserves	Total	Tanks	Fighting vehilces	Artillery	Ballistic missile launchers
Eastern Mediterranean							
Egypt	450,000	254,000	704,000	3,100	3,680	3,556	24
Israel	186,500	445,000	631,500	3,630	6,870	1,348	+
Jordan	100,700	60,000	160,700	920	1,773	871	
Lebanon	61,400		61,400	280	1,235	335	
Palestinian Authority	45,000		45,000				
Syria	289,000	132,500	421,500	3,700	5,060	3,274	~45
Turkey	421,000	379,000	800,000	2,600	5,885	4,370	12
Persian Gulf							
Bahrain	8,200		8,200	180	277	48	9
Iran	520,000	350,000	870,000	1,620	1,400	2,700	40
Iraq	9,750		9,750	20	120	0	0
Kuwait	15,500	24,000	39,500	293	672	127	
Oman	34,000		34,000	124	346	133	
Qatar	11,800		11,800	30	280	56	
Saudi Arabia	171,500	20,000	191,500	750	4,430	410	12
UAE	65,500		65,500	532	1,044	337	6
North Africa and others							
Algeria	127,000	150,000	277,000	900	1,915	920	
Libya	76,000		76,000	650	2,230	2,320	~80
Morocco	198,500	150,000	348,500	285	1,089	1,060	
Sudan	104,000		104,000	350	575	778	
Tunisia	35,500		35,500	139	326	205	
Yemen	65,000	200,000	265,000	715	495	675	10

The Middle East Military Balance at a Glance (continued)

Air Force			Air Defense			Navy		
Combat aircraft	Transport aircraft	Helicopters	Heavy batteries	Medium batteries	Light launchers	Submarines	Combat vessels	Patrol craft
Eastern Mediterranean								
505	47	230	109	44	105	4	59	103
470	70	181	23		70	3	15	40
96	14	83	17	17	50			10
		24						20
350	23	195	108	64	55		16	8
400	90	467	24		86	12	83	106
Persian Gulf								
33	3	48	1	2	40		11	22
203	80	340	30		95	3	56	160
0	3	6	0		0		0	5
39	5	25	12	1			10	77
29	50	51			58		9	68
12	7	24			51		7	13
256	38	214	25	21			27	68
93	30	102	5	6	115		12	134
North Africa and others								
213	41	174	11	18	78	2	26	16
290	72	112	30	10	55		24	
59	41	122			37		15	52
40	14	46	20					16
18	15	47			83		15	40
52	14	24	25		120		10	8

Weapons of Mass Destruction

State	Chemical	Biological	Nuclear	SSM Launchers		
				Up to 150 km.	150–600 km.	600–3,000 km.
Eastern Mediterranean						
Egypt	weapons program	R&D	R&D		24	+
Israel	R&D	R&D	alleged weapons	12		+
Syria	weapons program	weapons program	R&D	18	26	1
Turkey	none	none	R&D	12		
Persian Gulf						
Bahrain	none	none	none	9		
Iran	weapons program	weapons program	R&D	16	20	5
Saudi Arabia	none	none	none			12
UAE	none	none	none		6	
North Africa and others						
Algeria	none	none	R&D			
Libya	none	none	none		80	
Sudan	weapons program	none	none			

Space Assets

State	Imagery ground stations	Communication satellites	Research satellites	Reconnaissance satellites	SLVs
Eastern Mediterranean					
Egypt	+				
Israel		+	+	+	+
Syria	+				
Turkey		+	+		
Persian Gulf					
Saudi Arabia	+	+			
UAE	+	+			
Iran			+		+
North Africa and others					
Algeria		+	+		

The Eastern Mediterranean Military Forces

Eastern Mediterranean – Personnel

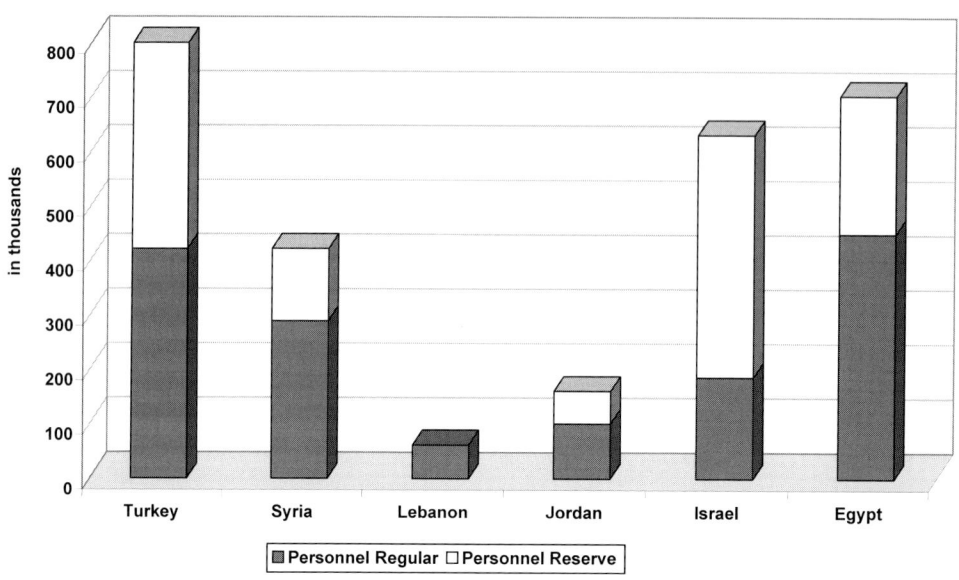

Eastern Mediterranean – Armor

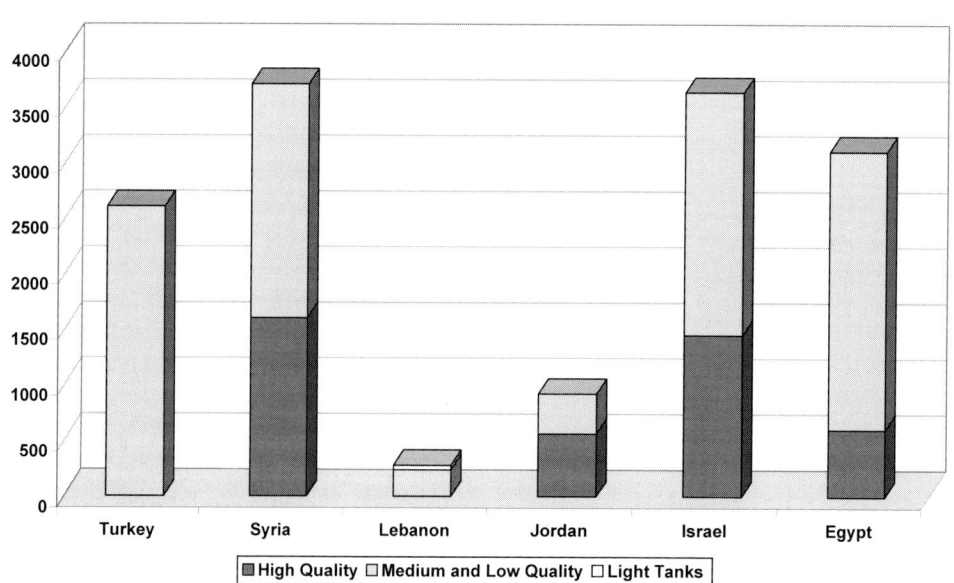

The Eastern Mediterranean Military Forces (continued)

Eastern Mediterranean – ACVs

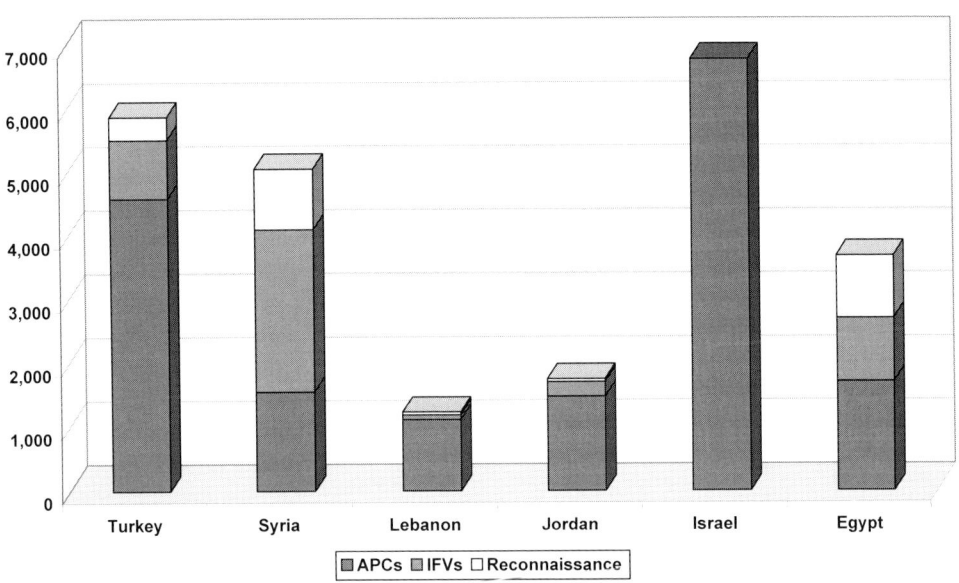

Eastern Mediterranean – Artillery

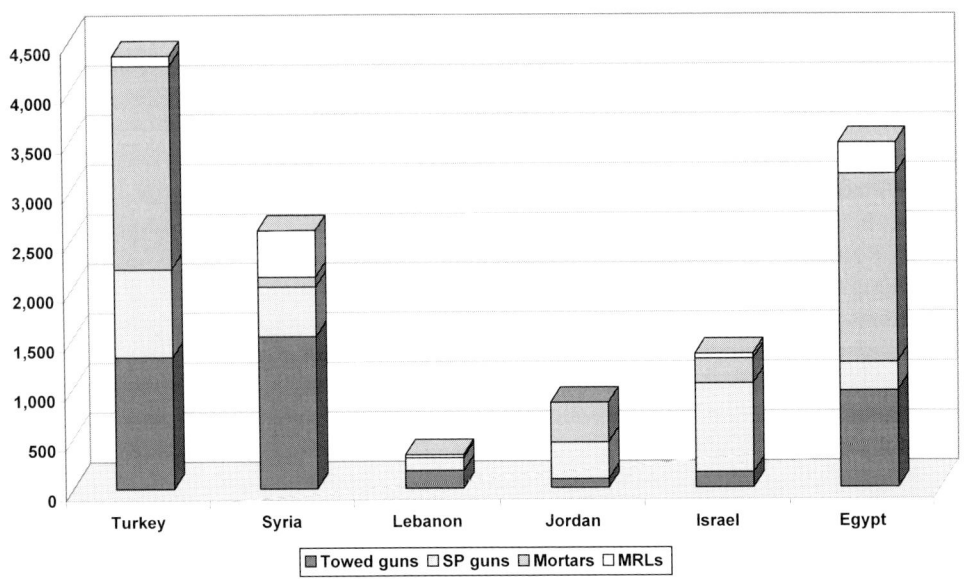

The Eastern Mediterranean Military Forces (continued)

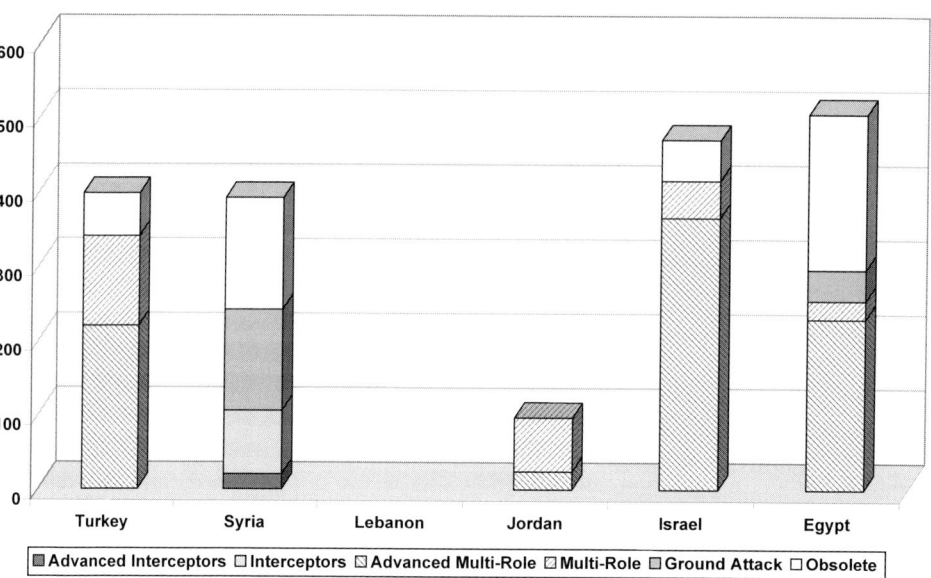

Eastern Mediterranean – Combat Aircraft

■Advanced Interceptors □Interceptors ▨Advanced Multi-Role ▨Multi-Role ▨Ground Attack □Obsolete

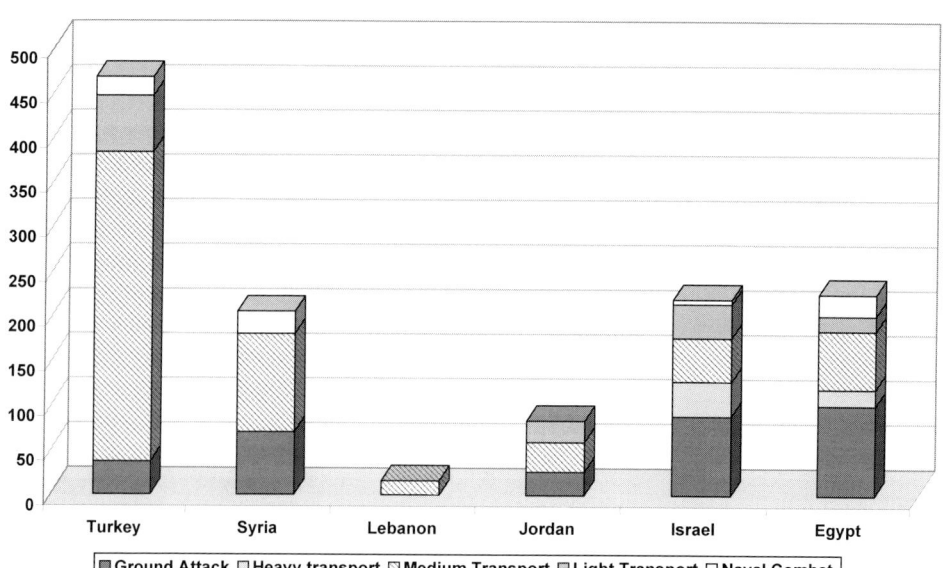

Eastern Mediterranean – Helicopters

▨Ground Attack □Heavy transport ▨Medium Transport ▨Light Transport □Naval Combat

The Eastern Mediterranean Military Forces (continued)

Eastern Mediterranean – Air Defense

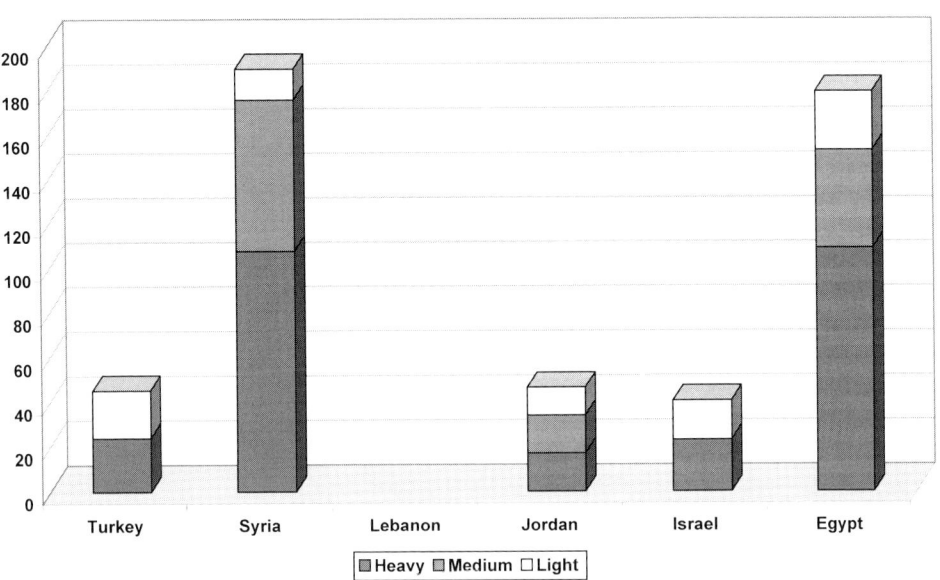

Eastern Mediterranean – Naval Combat Vessels

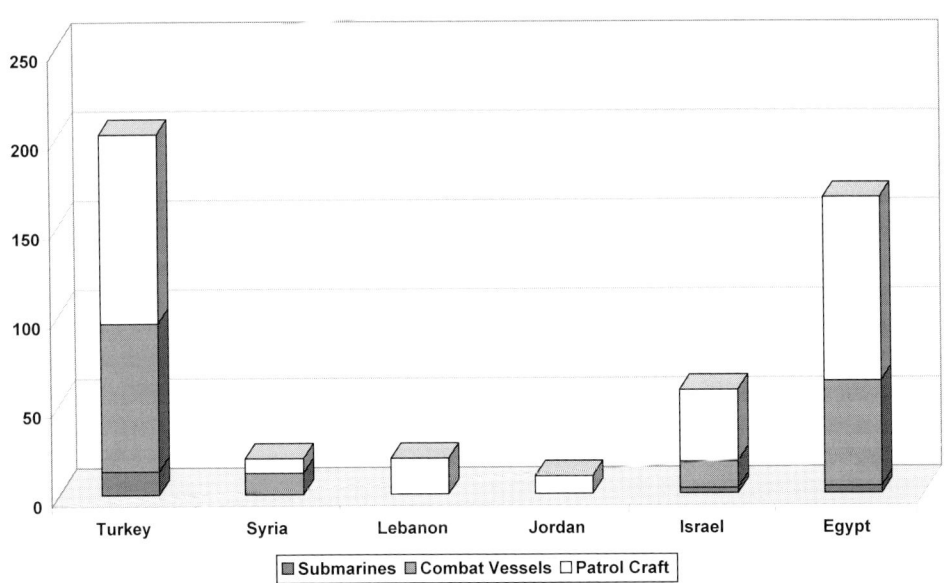

The Persian Gulf Military Forces

The Gulf – Personnel

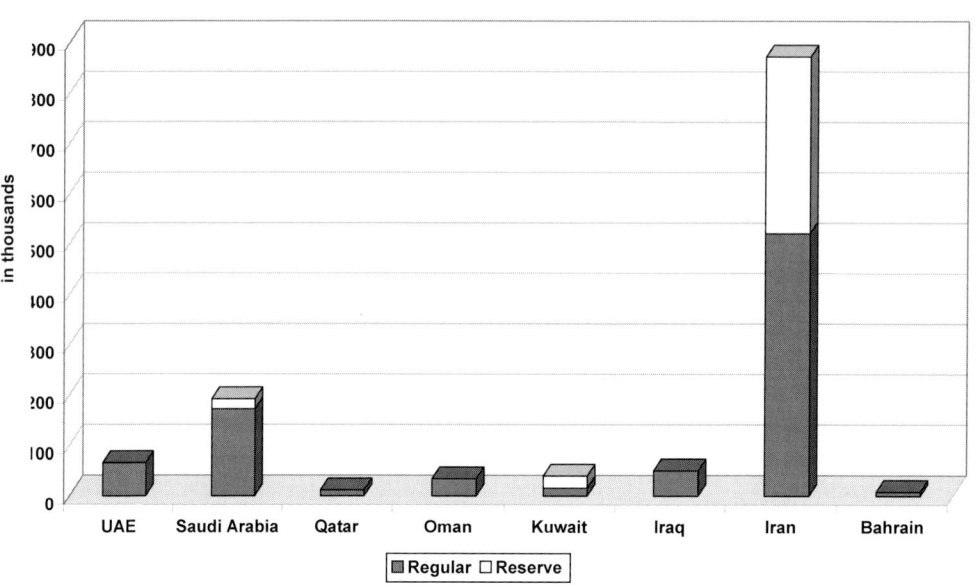

The Gulf – Armor

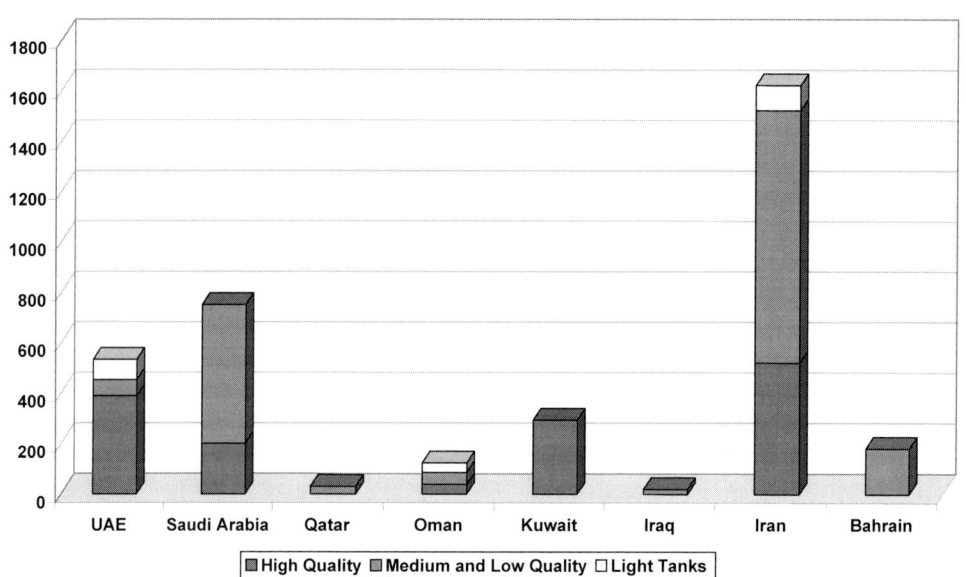

The Persian Gulf Military Forces (continued)

The Gulf – ACVs

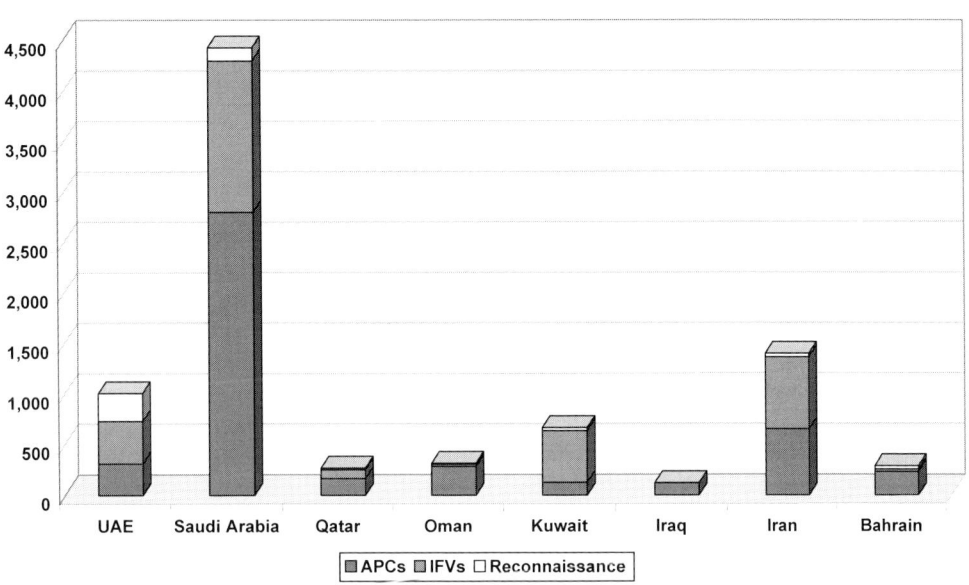

APCs ■IFVs □Reconnaissance

The Gulf – Artillery

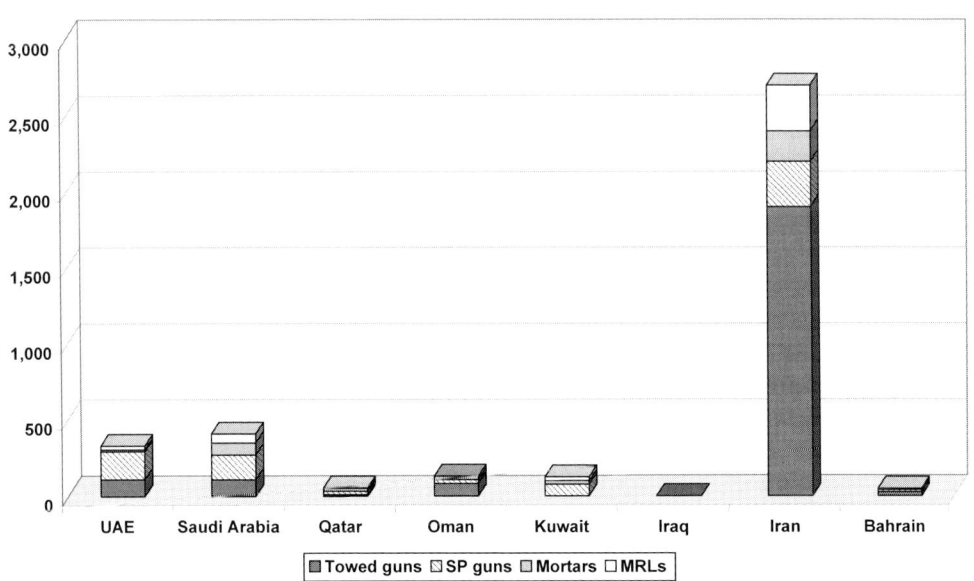

■Towed guns □SP guns ■Mortars □MRLs

The Persian Gulf Military Forces (continued)

The Gulf – Combat Aircraft

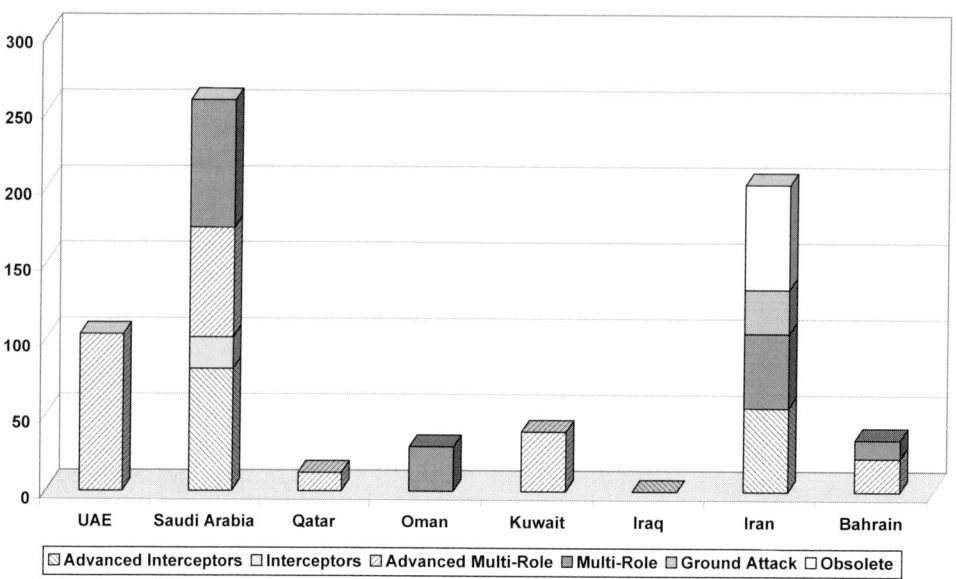

The Gulf – Helicopters

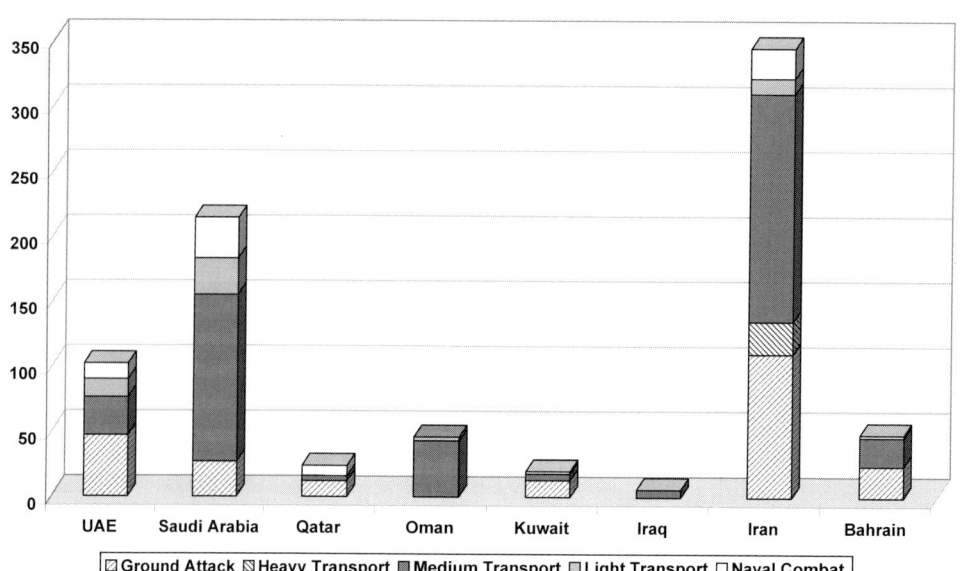

The Persian Gulf Military Forces (continued)

The Gulf – Air Defense

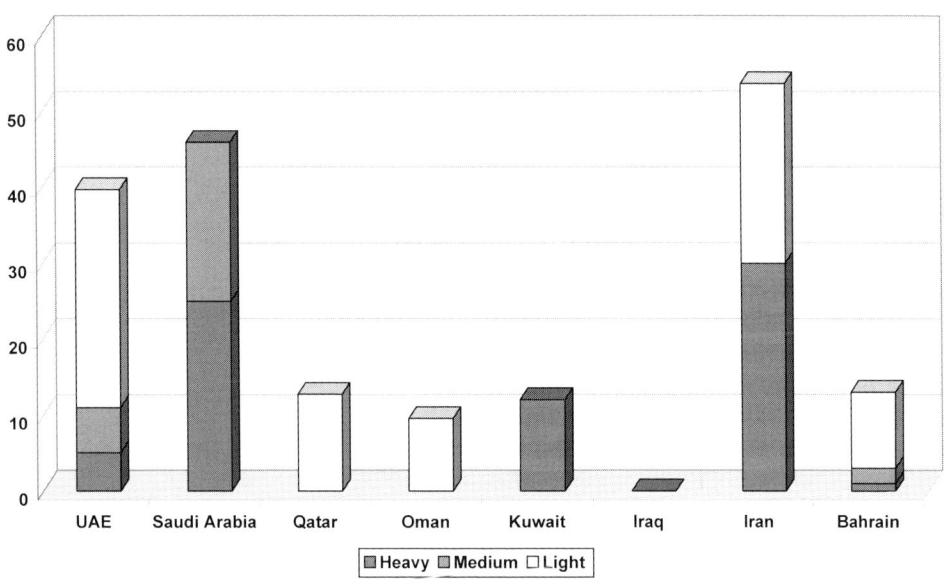

The Gulf – Naval Vessels

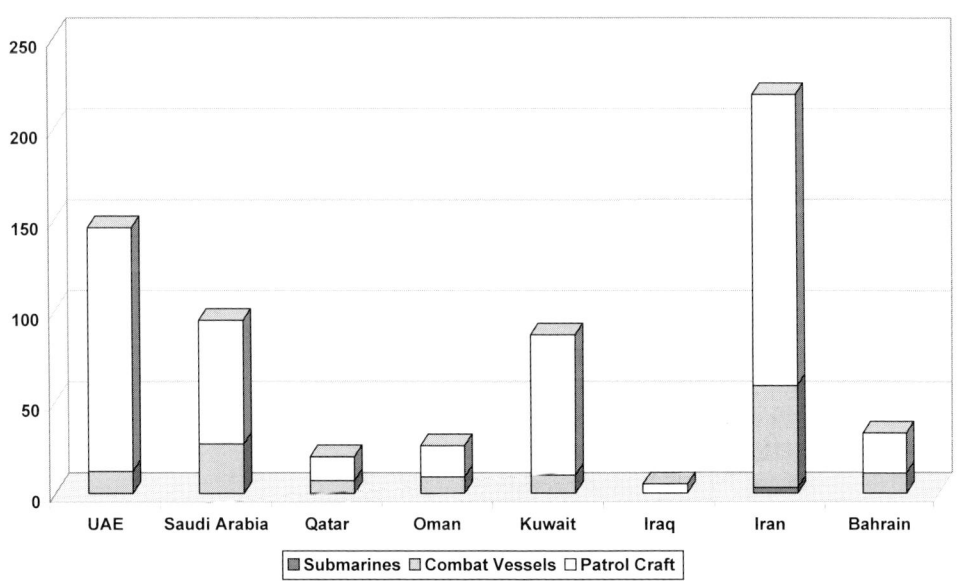

The North African Military Forces

North Africa – Personnel

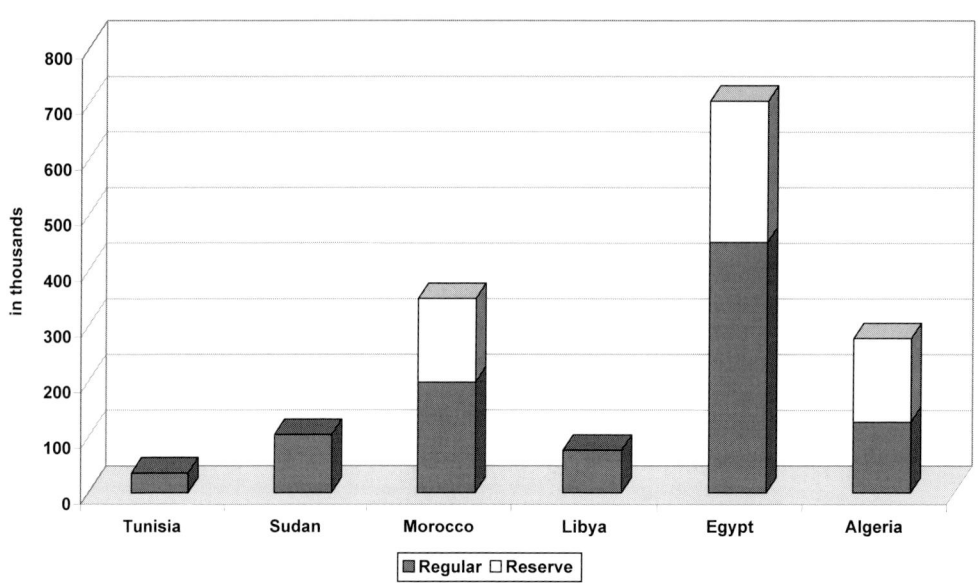

North Africa – Armor

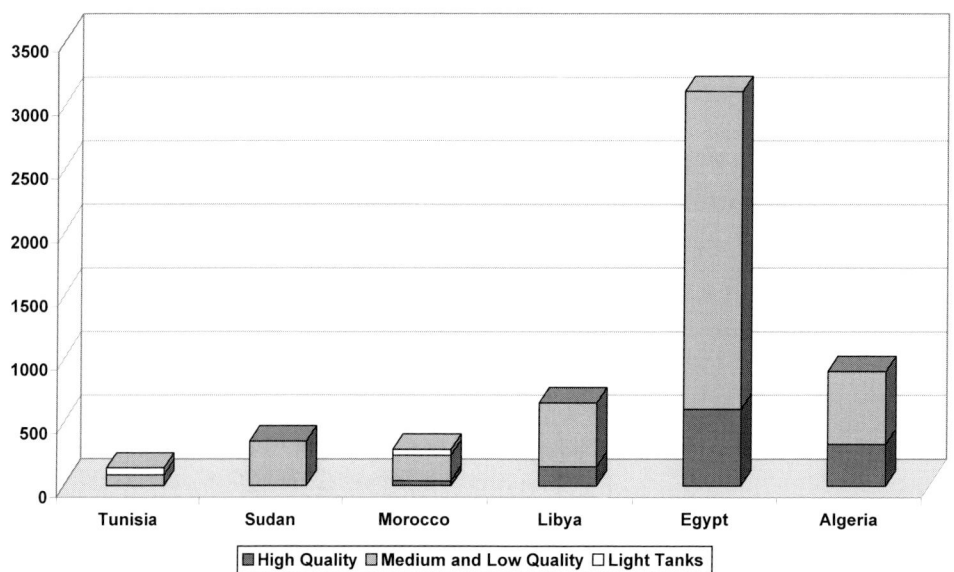

The North African Military Forces (continued)

North Africa – ACVs

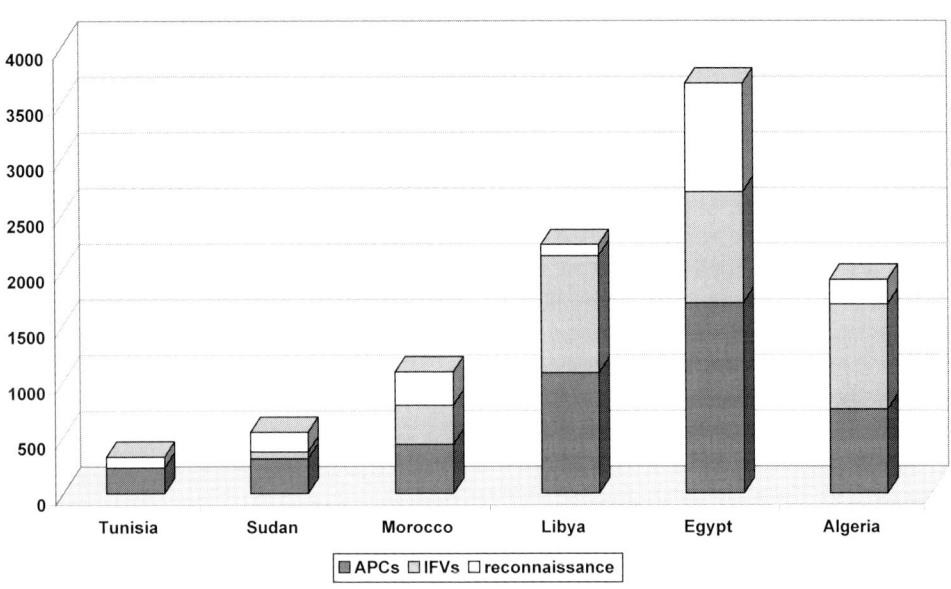

North Africa – Artillery

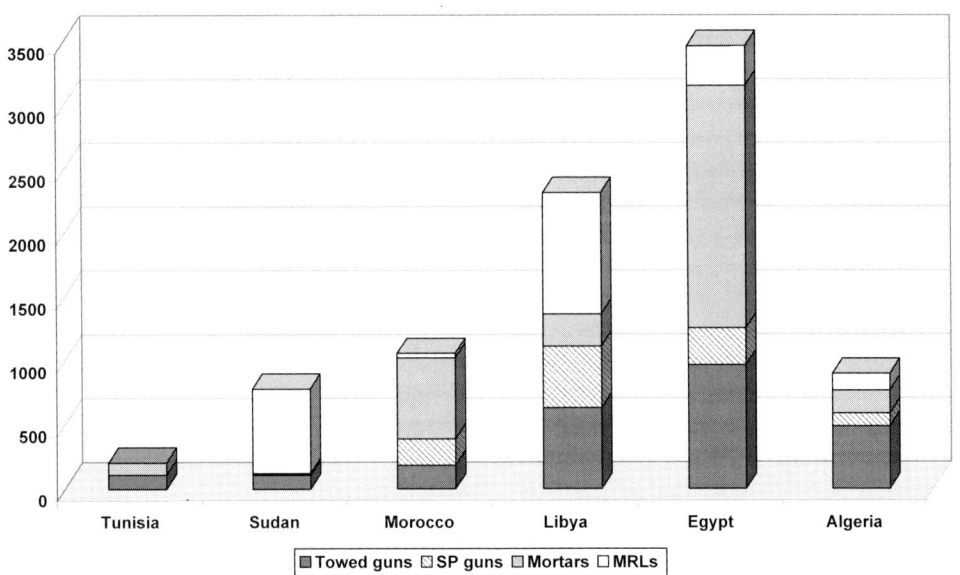

The North African Military Forces (continued)

North Africa – Combat Aircraft

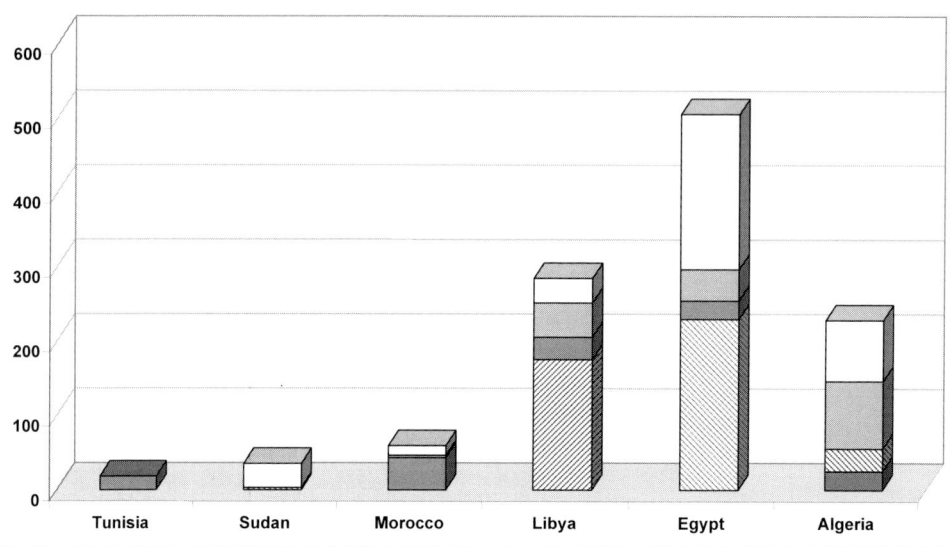

North Africa – Helicopters

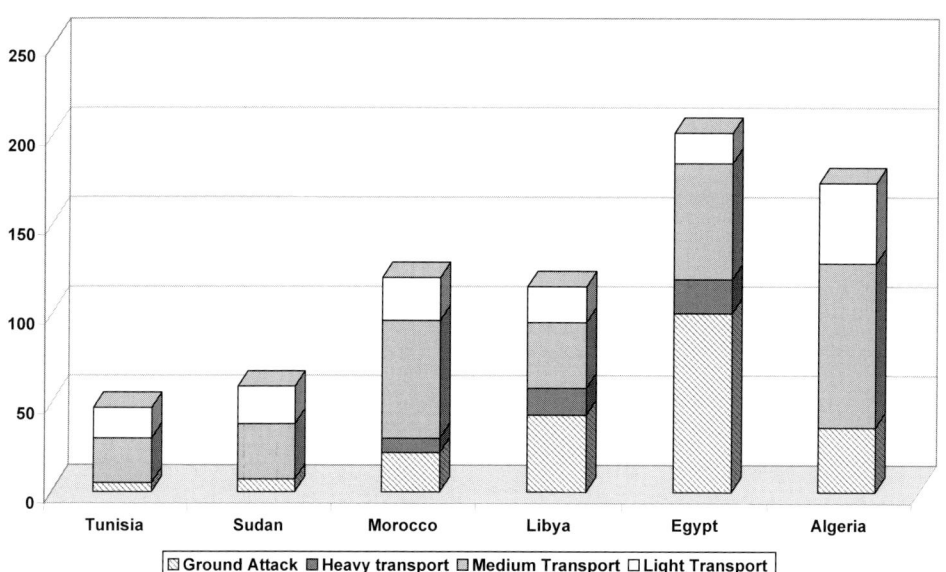

The North African Military Forces (continued)

North Africa – Air Defense

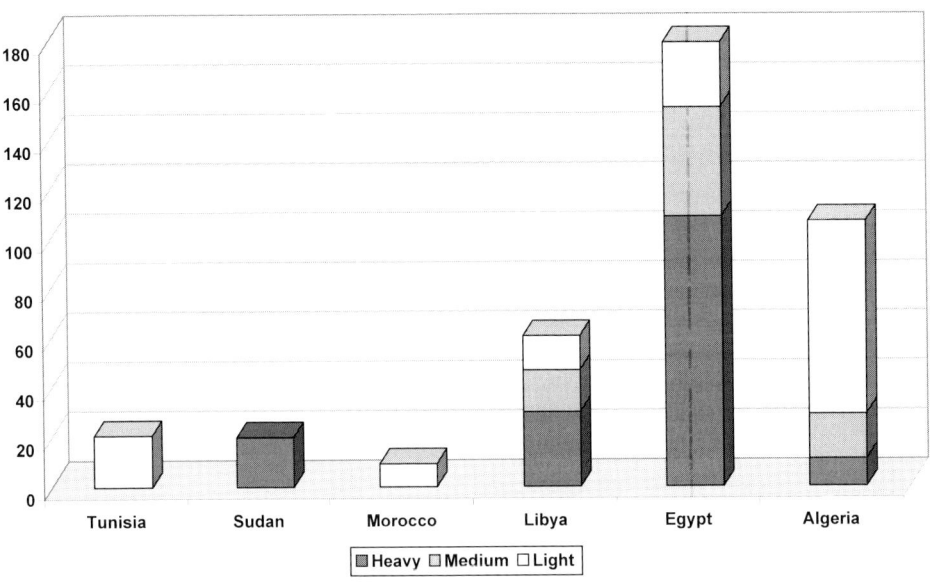

North Africa – Naval Vessels

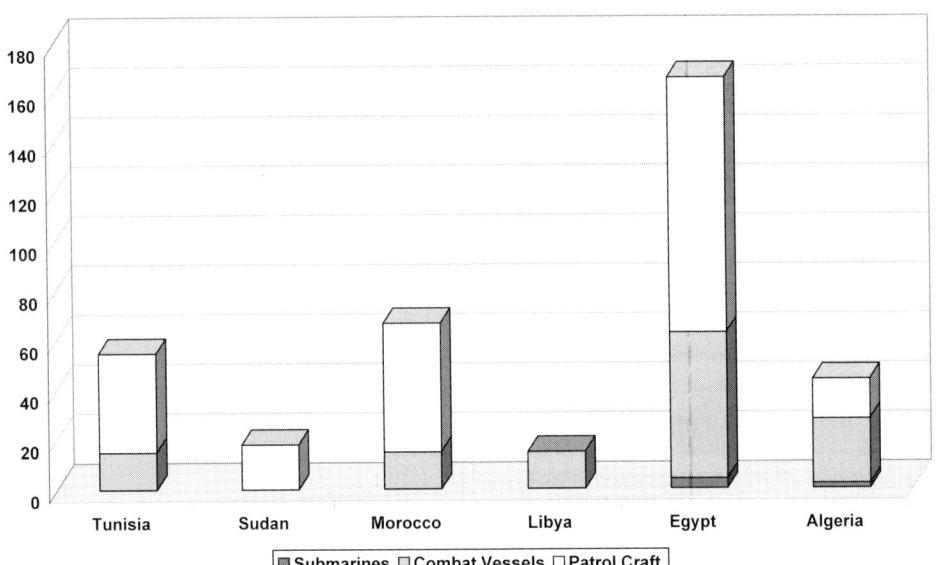

Contributors

Editors

Zvi Shtauber was appointed Head of the Jaffee Center in February 2005, following a long career in the Israel Defense Forces and other public offices. His final position in the military was head of the IDF Strategic Planning Division. Prior to that he served as the assistant defense attaché in the Israeli embassy in Washington. Upon his retirement from the IDF, Dr. Shtauber served as vice president of Ben-Gurion University of the Negev. In 2001 he was appointed Israeli ambassador to the Court of St. James (UK), a position he held until mid-2004. Dr. Shtauber was a member of the Israeli delegations in peace talks between Israel and its neighbors, including the talks with Syria at Shepherdstown and with the Palestinians at Camp David. He also represented the IDF in multi-lateral negotiations on regional security.

Yiftah S. Shapir joined the Jaffee Center in 1993 as an associate of the Center's Project on Security and Arms Control, where he followed the proliferation of weapons of mass destruction (WMD) in the Middle East. He is head of JCSS's Middle East Military Balance Project, and he is responsible for the quantitative section of the annual *Middle East Strategic Balance.* Shapir served as an officer in the Israeli Air Force, and has extensive background in information technology and operations research.

Other Contributors

Shlomo Brom joined the Jaffee Center as a Senior Research Associate in 1998 after a long career in the IDF. His most senior post in the IDF was Head of the Strategic Planning Division in the Planning Branch of the General Staff. Brig. Gen. Brom participated actively in peace negotiations with the Palestinians, Jordan, and Syria. In 2000 he was named Deputy to the National Security Advisor, returning to JCSS at the end of his post. He is the director of the Jaffee Center's Israel Defense Policy Review project.

Mark A. Heller is Director of Research at the Jaffee Center and editor of *Tel Aviv Notes.* He has been affiliated with the Jaffee Center since 1979 and has taught international relations at Tel Aviv University and at leading universities in the US. Dr. Heller has written extensively on Middle Eastern political and strategic issues. He is also currently a member of the Steering Committee of EuroMeSCo, the Euro-Mediterranean consortium of foreign policy research institutes.

Ephraim Kam, Deputy Head of the Jaffee Center, served as a colonel in the Research Division of IDF Military Intelligence until 1993, when he joined the Jaffee Center. Positions he held in the IDF included Assistant Director of the Research Division for Evaluation and Senior Instructor at the IDF's National Defense College. Dr. Kam specializes in security problems of the Middle East, strategic intelligence, and Israel's national security issues.

Emily B. Landau is director of the Jaffee Center's Arms Control and Regional Security Project at JCSS. She has published on CSBMs in the Middle East, Arab perceptions of Israel's qualitative edge, Israeli–Egyptian relations, Israel's arms control policy, and the Arms Control and Regional Security working group of the Madrid peace process (ACRS). Dr. Landau's current research focuses on regional dynamics and processes in the Middle East and developments in arms control thinking.

Paul Rivlin has a joint appointment at the Jaffee Center and at the Moshe Dayan Center for Middle East and African Studies at Tel Aviv University. Dr. Rivlin specializes in the political economies of Arab states, and has published widely on defense economics, oil market trends, and economic development in the Middle East.

Yoram Schweitzer is an expert on international terror at the Jaffee Center. He has lectured and published widely on terror-related issues, and consults for government ministries on a private basis. His areas of expertise include the "Afghan alumni" phenomenon, al-Qaeda, the internationalization of suicide terrorism, and state-sponsored terrorism.